The Management of Safety

The behavioural approach to changing organizations

Valerie J. Sutherland,
Peter J. Makin and Charles J. Cox

SAGE Publications
London • Thousand Oaks • New Delhi

First published 2000

SAGE Publications Ltd
6 Bonhill Street
London EC2A 4PU

SAGE Publications Inc
2455 Teller Road
Thousand Oaks, California 91320

SAGE Publications India Pvt Ltd
32, M-Block Market
Greater Kailash – I
New Delhi 110 048

British Library Cataloguing in Publication Data

A catalogue record for this book is
available from the British Library

ISBN 0 7619 6612 9
ISBN 0 7619 6613 7 (pbk)

Library of Congress catalog card number available

Typeset by Keystroke, Jacaranda Lodge, Wolverhampton
Printed and bound in Great Britain by Athenaeum Press,
Gateshead

CONTENTS

PREFACE

A day or two after the Piper Alpha disaster, the three of us were having coffee in the common room of the Manchester School of Management at UMIST. The course of our conversation led us into a consideration of why, despite the great emphasis placed on safety and accident prevention, particularly in the off-shore oil industry, these major accidents continued to happen. With our interest and background in the application of behaviour modification in organizations it was, perhaps, inevitable that we should start to analyse these incidents by looking for what was reinforcing the behaviour that was leading to such accidents. According to the theory, if accidents continue to happen, this must be due to continuing unsafe behaviour and something must be rewarding that behaviour for it to continue. It then seemed clear that the organization or the 'system' must, in some way, be rewarding unsafe behaviour. Closer examination showed that, not only was this the case, but that in some instances the organization was actually inadvertently punishing the behaviour that was needed if safety was to improve. That organizations have this potential to punish the very behaviour that they actually want, and reward what they do not want, is discussed in detail in Chapter 1, and developed further in Chapter 3. If safety is to be improved, this pattern of reinforcement must, somehow, be reversed.

At this stage the discussion led to two of the present authors writing a paper on this phenomenon for the journal of the Royal Society for the Prevention of Accidents (Makin and Sutherland, 1991). This was seen by the manager responsible for safety at a major cellophane manufacturer, who wrote to us inviting us to try out try these ideas in his plant. This we did with considerable success, as described in Chapter 4. From this initial successful intervention we have moved on to apply the process in many other organizations across a number of industries. We are also starting to apply the technique to other aspects of organizational behaviour, such as quality improvement (see Chapter 7). We have also, of course, steadily developed the basic model on which the approach is based. We have now written this book to explain the underlying theory, and the techniques for application, of the behavioural approach to safety so that others may apply it in their own organizations. It is our hope that it will be useful to managers, safety specialists and students on management programmes.

1

THE NEED FOR CHANGE

Many organizations endeavour to improve workplace safety performance through both task and hardware approaches. Typically, in the past, these have focused on physical design and its relationship to employee performance and so engineers and ergonomists have dominated the safety arena. More recently they have been joined by sociologists and psychologists who try to explain accident involvement and poor safety performance in terms of either attitudes, safety climate and/or organizational culture. The first of these approaches involves attempting to change employee attitudes and reduce accidents through the use of publicity campaigns, safety training or disciplinary actions. It can be recognized as the 'traditional' approach to improving safety performance. Other approaches tend to adopt the method of workplace survey (that is, interview and questionnaire techniques) with the aim of diagnosing problems.

Although each of these has a part to play in any safe work environment and each has offered relative degrees of success in improving safety in the workplace, for a variety of reasons they have not had the anticipated impact in reducing accident rates at work. So, the challenge to continually improve safety performance remains a key activity for many organizations. Indeed, managers and safety directors responsible for improving safety performance have contacted us asking for help because they believe that they have reached some sort of plateau. They have felt unable to drive down the level of accidents any further by the use of traditional means. Now, a growing body of evidence suggests that a shift in focus from traditional approaches to a *behaviour-based approach to safety improvement* can further improve workplace safety performance and will reduce the number and the severity of occupational accidents.

In this chapter we consider the need for change in the approach to safety management and offer a rationale for the introduction of a behaviour-based approach to safety improvement, in the light of the weaknesses inherent in traditional safety improvement techniques. Whilst we suggest a need for organizations to change if they are to continue to improve workplace safety, we are not saying that traditional approaches to safety in the workplace are wrong, only that they do not satisfy the demands on safety management in contemporary organizations. Neither are we saying that the behaviour-based approach to safety improvement will replace the initiatives, systems and practices currently in place. It simply offers another option towards a more integrated strategy for safety improvement and one which meets the needs of people at work and the prosperity of the organization.

We begin with an examination of accident statistics to review the size of the problem.

The High Cost of Ineffective Safety Management

Failure to manage safety effectively in the workplace is evident and the penalties are high. Each year, it is estimated that around £11–16 billion will be lost to British business through occupational accidents and illness. According to the Royal Society for the Prevention of Accidents (1997), every working day in Britain there are approximately:

- 2 accidental deaths in the workplace
- 2,500 reportable injuries
- 9,000 work injuries
- 250,000 thousand 'near misses'

and

- 7 out of 10 occupational accidents (and cases of ill health) could easily have been prevented
- over 2,000 lives a year could have been saved if health and safety laws were followed by employers AND employees.

Therefore, the occupational safety and welfare of the workforce is of paramount importance to the functioning and profitability of an organization. Whilst a lack of attention to safety matters may incur significant costs to the individual, the organization and society, the full loss is often not understood or correctly calculated because there may be direct and indirect costs such as:

- lost production caused by:
  - time away from job by injured person and co-workers in attendance
  - time spent by first aider attending injured person
  - time spent investigating the accident, attending hearings, completing paperwork, by the supervisor, manager, safety representative and/or union representative
  - possible down-time of production process
  - possible damage to product, plant and equipment

- time and costs due to repair of plant and equipment
- increased insurance premiums
- fines/legal costs
- medical expenses

- compensation costs to injured employee
- absence from work
- costs of a temporary replacement, or the costs of strain on the employees who are left to cope with a reduced manpower level
- finding alternative work for an employee who is unable to fulfil previous job commitments
- loss of investment in employee development when the individual is forced to retire early
- cost to replace and re-train a new member of staff
- loss of experience to the organization
- lowered morale of both victim and work colleagues; this often has an adverse impact on motivation levels, leading to poor performance and productivity
- potential for unsatisfactory employee relations
- media attention, bad publicity and poor public relations
- inability to attract new recruits and/or retain staff (that is, all costs associated with a high labour turnover).

Likewise, rarely are the *full* costs of accident involvement for the victim acknowledged. These direct and often hidden costs might be in terms of:

- pain and suffering to the injured
- the overspill impact and effect on family and dependants
- absence from work and the subsequent isolation from colleagues and friends
- loss of income (e.g. overtime bonus etc.)
- physical or psychological disability forcing a change in job status
- reaction of others to disability
- forced early retirement
- change in lifestyle (leisure, interests and hobbies)
- the 'never quite the same again' implications, and/or long-term physical and psychological health problems.

Thus, it can be said that we probably only consider the 'tip of the iceberg' when we attempt to place a cost on industrial accidents. There are many other costs 'under the surface'.

A Need to Reduce the Level of Accidents

Many organizations have driven the level of workplace and work-related accidents down to very low levels, but recent statistics suggest a current overall increase in accident levels. The number of injuries at work in the year to April 1997 has reached the highest level since 1989/90 (see Table 1.1). Fatality totals and rates rose, after falling for more than seven successive years; and as can be seen in Table 1.2, major injuries more than doubled

Table 1.1 *Injuries to employees, self-employed and the public, 1989–1997*

	1989/90	1990/91	1991/92	1992/93	1993/94	1994/95	1995/96	1996/97*
Fatal	681†	572	473	452	403	376	344	696
Major	33,084	31,203	29,707	28,722	29,531	30,996	30,968	65,402
Over 3-day	167,109	162,888	154,338	143,283	137,459	142,218	132,976	129,491
Total	200,874	194,663	184,518	172,457	167,393	173,590	164,288	195,589

* Provisional.
† Includes 95 Hillsborough Stadium fatalities

Source: Health and Safety Commission (1997) *Plan of Work 1997/98*; and Health and Safety Commission (1996) *Annual Report 1995/96*

Crown copyright material is reproduced with permission of the Controller of Her Majesty's Stationery Office.

from the year 1995/96, particularly in manufacturing. Also accidents to the self-employed (particularly in agriculture and construction) dramatically increased (Health and Safety Commission, *Annual Report* 1996/97). Although some of the increases might be attributed to changes introduced by the *Reporting of Injuries, Diseases and Dangerous Occurrences Regulations 1995* (RIDDOR, 1995), there is cause for concern and disappointment about this apparent deterioration in safety performance.

Whilst there is always likely to be some dispute about official statistics, the figures in Tables 1.1 and 1.2 are sobering. It remains a fact that too many people are still being killed or injured in some way as they engage in day-to-day work activities. Without a doubt there exists a need to find additional ways to improve safety. Indeed, it is also a commonly expressed notion (by those at the 'sharp end') that not all accidents are reported because traditional safety systems and processes (for example, industry league tables) actively discourage the employee from coming forward and owning up to accident involvement, even though policy and mission statements would seem to imply the opposite. A desire to apportion blame and punishment remain deterrents to a safe work environment. In fact, it is common practice for the organization to inadvertently reward the individual for non-reporting and cover-up (the theoretical basis for this is explained in Chapter 3). These circumstances serve to reinforce a culture and climate which lacks the openness and trust required if one is to learn anything useful from accidents, dangerous occurrences or the near miss events that occur.

Learning from the Past – a Flawed Strategy!

It is our contention that current safety systems are not, despite the good intentions behind them, producing the required results. These systems often encourage the 'wrong' behaviour in those who are likely to be the victims

Table 1.2 *Employee injury totals and rates 1995/96 (final) and 1996/97 (provisional) by sector*

Sector	1995/96							1996/97						
	Injury numbers				Rates per 100,000			Injury numbers				Rates per 100,000		
	Fatal	Major	Over 3-day	Total	Fatal	Major	Over 3-day	Fatal	Major	Over 3-day	Total	Fatal	Major	Over 3-day
Agricultural/forestry/fishing	20	408	1,279	1,707	7.8	158.6	497.3	20	677	1,473	2,170	7.9	268.4	583.6
Energy and water supply	18	508	3,173	3,699	8.0	225.9	1,411.5	8	685	2,862	3,555	4.3	364.9	1,523.8
Manufacturing	42	5,146	42,097	47,285	0.9	130.5	1,067.4	57	8,220	39,855	48,132	1.4	209	1,013.4
Construction	62	1,806	8,305	10,173	7.7	222.9	1,030.3	68	3,239	9,055	12,362	8.4	397.8	1,112.0
Services	67	8,110	72,465	80,642	0.4	50.1	447.5	66	15,219	73,974	89,259	0.4	91.9	446.6
Unclassified		590	3,263	3,853										
All industry	209	16,568	130,582	147,359	1.0	77.1	607.4	219	28,040	127,219	155,478	1.0	128.9	584.9

Source: Compiled from Health and Safety Commission (1997) *Plan of Work 1997/98*

of accidents. In order to demonstrate where many current systems are breaking down, descriptions of some incidents that occurred during the 1980s and 1990s follow. These provide clues to the weakness inherent in traditional approaches to safety in the workplace.

1 In a work climate characterized by change and downsizing, we find a workforce unfamiliar with the workplace, with a poor communications structure, often with limited training and undefined lines of responsibility. This was the likely setting for the following example:

A fire on an escalator at King's Cross Underground Station on the evening of 18 November 1987, which caused the death of 31 people. At 1930 hours a passenger reported a small fire on an escalator in the underground station. A member of staff who was normally based at another station went to investigate the problem but did not inform either the station manager or the line controller. A member of the Metropolitan Police called the fire brigade and started to evacuate passengers. London Fire Brigade personnel reached the ticket hall at approximately 1943 hours, but two minutes later the fire erupted and the resulting flashover engulfed the ticket hall as passengers were trying to escape. 'It was too late for them [the fire brigade] to do anything. Between 1930 and 1945 hours, not one single drop of water had been applied to the fire . . .' (Department of Transport Investigations Report: Fennell, 1998). Of London Underground staff it was said, 'while the actions of individuals at the time were understandable, and in several cases involved presence of mind and courage, their overall response may be characterised as uncoordinated, haphazard and untrained', and 'there seems to have been a lack of perception that fire demands a very rapid reaction if it is to be contained' (Fennell, 1998).

---

2 Traditional safety management systems place the responsibility for safety on managers and, perhaps, the safety committee. However, a 'right-sized' management team often complain about their limited opportunities for being 'at the sharp end'. Neither can they be omnipresent and 'police' a safety environment (it is doubtful that this would be tolerated anyway!). Also, the safety committee often lacks the power and authority needed to make a real difference. So, consider the safety limitations in the next two scenarios:

An explosion, heard up to 30 miles away, and resulting fires at the Texaco refinery site, Milford Haven, on 24 July 1994, causing injury to 26 people and massive damage to the site. The incident happened at the Pembroke Cracking Company (PCC) plant, a joint partnership project between Texaco Limited and Gulf Oil (GB), following a severe electrical storm, which precipitated plant disturbances. The Health and Safety Executive found that the direct cause of the explosion was a combination of failures which led to the release of 20 tonnes of flammable hydrocarbons from an outlet pipe which eventually ignited and exploded causing both major and secondary fires. It was found that this situation arose because of a combination of management, equipment and control systems failures, such as:

- attempts to keep the unit running when it should have been shut down
- a control valve being shut when the control system indicated it was open
- a modification had been carried out without assessing all the consequences
- control panel graphics did not provide necessary process overviews.

Even though an act of nature, a lightning strike, initiated the event, investigations indicated that the subsequent explosion was avoidable and the two companies were fined a total of £200,000 and order to pay £143,700 costs.

The sinking of the P&O Ferry *Herald of Free Enterprise* on 6 March 1987, as she sailed from Zeebrugge in Belgium with her bow doors open and capsized, killing 188 people. There was a high water spring tide. The ship was not designed for a port with only one ramp. At 1806 hours the ship reversed out. The Bosun saw the doors open but it was not his job to close them. This was the responsibility of the Chief Officer, who was required to be on the bridge 15 minutes before sailing. As the Master increased speed, water flowed in through the open doors. At 1827 hours, the ship capsized.

---

3   The traditional safety improvement model suggests that we investigate all incidents in order to learn to avoid costly and fatal mistakes. However, the following example indicates that we do not appear to learn from the past even when the incident is on the scale of a major disaster. Whilst acknowledging that all incidents must be investigated, we doubt the efficacy of this as a safety improvement strategy. It takes too long for the message to get through to other contexts. For example:

The sinking of the *Estonia* **in a storm off Finland,** 29 September 1994, costing 780 lives. It would seem that the faulty seals on the bow doors may have cost this considerable loss of life at sea because the lessons from the *Herald of Free Enterprise* appear to have gone unlearned. Despite earlier recommendations, only after this event were doors on ferries of this design welded shut.

4   Organizations devise and install complex systems which in practice are ignored or by-passed in order to meet the needs of production. Thus the constant message that 'safety comes first' is simply viewed as 'lip service' to safety. Everyone joins in this conspiracy of silence. The motives for this are explained in Chapter 3. An example of this can be seen in the events surrounding a major tragedy in the offshore oil and gas exploration and production industry:

The loss of the North Sea drilling rig *Piper Alpha* on the night of 6 July 1988. This devastating event appears to be the culmination of a catalogue of accidents in the quest for oil over the preceding decades. Caused by a massive explosion, it claimed the lives of 167 people and the total loss of the rig *Piper Alpha*. The subsequent investigation and public inquiry, completed in 1990, chaired by Lord Cullen, produced a list of over 100 recommendations for the industry. Quite clearly the behaviours of personnel on the rig (including the offshore installation managers' failure to cope with the problems they faced on the night of the disaster) were contributory factors in this catastrophe. The Cullen report is perhaps the most comprehensive study of safety in the North Sea and has provided psychologists with yet further examples, if they were needed, of how organizations ignore human factors to their own, and often the public's, cost.

Indeed, long after the publication of the Cullen Report the debate still continues because of the more recent findings from Lord Caplan's inquiry (which began in 1993 when Occidental Petroleum (Caledonian) Limited, Occidental's successor, sued seven contractors in a bid to recover compensation paid out to the families of the 167 people who died). Both families and trade unions are still very concerned because the blame for events which triggered this explosion has been placed on two of the deceased workers who were, at the time, carrying out repair work on a pump. It is suggested that one of the men failed to tighten bolts on the blind flange. Whilst Lord Caplan attributed the primary cause of the accident to the two men, he also held the oil rig operator partly responsible because, 'their arrangements for monitoring and auditing the implementation of these [safety] procedures left much to be desired'.

Not surprisingly, there is a rising tide of anger at these new allegations since the men are not alive to explain their actions and possibly exonerate themselves . . . and, the simple act of tightening or not tightening bolts could hardly be conceived to lead to a disaster of such damaging proportions! The original verdict of Lord Cullen was that the two men had no way of knowing that their actions were unsafe because vital paperwork had not been passed on from the previous shift. It is suggested that had they received this paperwork it would have been obvious to them that they should not have continued with their actions. Lord Cullen judged that the permit to work system was dysfunctional on the night of the disaster, as it had been on previous occasions. Indeed, the permit to work system used by OPCAL had been found to be faulty before the fateful night in 1988, resulting in warnings for Occidental, from the regulatory authorities, to improve its safety performance. (In 1984, an uncontrolled release of hydrocarbons resulted in the evacuation of *Piper Alpha* and in 1986 a rigger was killed.)

A Fatal Inversion – Organizations Continue to Reward the Wrong Behaviour!

Many questions emerge from the foregoing events and we might ask why people continue to carry on working, whilst accepting that they are operating with faulty systems and equipment. Why might someone commit an act of behaviour which has potential to be the trigger for an accident or disaster? Indeed, why do accidents continue to happen despite the millions of pounds that companies spend trying to improve their record, and despite the expressed desire of top management to take steps to improve it? The answer, we believe, lies in the way that company policies influence the way people behave. Despite the best of possible motives, safety systems often encourage the very behaviour that management would like to discourage. There are numerous examples of organizations doing precisely this. One common example is the allocation of annual budgets within organizations. Most organizations urge all their departments to be as efficient as possible and to save money. If, at the end of the financial year, an efficient department has heeded this advice and is under-spent on its budget, what happens? In our experience the surplus is often 'clawed back' by the financial administrators and the following year's budget is cut. The department has been punished for doing precisely what was asked of it! This and other examples will be considered more fully in Chapter 3.

When accidents are reported often much emphasis is put on blaming the individual. The pressures on the individual to accept the blame are considerable. Most obvious is the threat, real or imagined, to job retention. Secondly a small, but immediate, compensation pay-out from an insurance company is better than a possibly larger sum at some indefinite time in the future. In addition, if the worker is involved in impending litigation, no other company will employ such a 'trouble-maker'. The company is also rewarded for having the worker accept the blame. If the person themselves can be shown to have been negligent, then the company is 'off the hook'.

Management style may also contribute to the pressures of non-reporting. For example, drilling rigs and platforms are tough 'macho' work environments and this is reflected in the predominant management style. Indeed, this is not uncommon within the 'heavy' industries, which are dominated by an all male workforce. On a drilling rig you tend not to question orders. As one roughneck told us, 'it is bawl and shout . . . so you are scared to ask questions'. According to one technician, 'there are plenty of people to make bullets but few with the courage to fire them'. This same mechanic had resigned from the safety committee because the minutes were edited by management, no doubt because of their own fear of the consequences of 'bad' reports. The whole system, therefore, encourages silence, non-reporting and perhaps, even, 'cover-ups'. The 'fringe' paper *Blowout* is an exception to this, where safety lapses are reported, often without the source being named. The newspaper, of course, gets rewarded for reporting these through higher sales.

Although the example used has been taken from the oil industry, it is our belief that many organizations operate in a similar manner. It might also be thought that these criticisms apply to only 'major' incidents. This is not so! Many minor, but frequent, accidents may go unreported for similar reasons. Also, the 'hassle' of filling in forms following a minor accident is reported both by workers who have to fill them in, and management who have to follow them up. Thus it is immediately more rewarding simply to ignore the unsafe event and hope that nothing more serious develops. Since this tends to be the most usual outcome, these assumptions and decisions normally prove to be correct. However, sadly and all too often, a catalogue of minor events underlies some major and possibly fatal incident. The accident investigators find out, too late, that the situation could have been avoided.

How then can systems be changed so as to encourage higher levels of safety? Thus far, we have used a traditional model of safety improvement because we have only examined past events in order to learn from them. Now we wish to argue that we need to become more 'proactive' in our approach to safety improvement. We start this in the following section, which describes key differences between the traditional versus the behavioural approach to safety improvement.

Moving from a Traditional Approach to a Behavioural Approach to Continuous Safety Improvement

From a reactive to a proactive approach

As we have seen, the traditional approach to safety is mainly *reactive* and operates by investigating and learning from incidents as they occur. Considerable effort is expended in this activity, which includes an in-depth reporting and examination of all incidents, dangerous occurrences and near misses.

Whilst this strategy has considerable value and must always exist, it is incomplete and flawed because:

1 Someone has already suffered an injury. It is ethically wrong and morally unsound to wait until 'something happens' before taking action.
2 The attention of the workforce is focused on a negative event with the aim of bringing about positive change. Propaganda, in the form of accident statistics, or exhortations on notice boards is rarely effective because:

   • negative information, for example notice boards which simply declare 'X number of days since last accident', does not indicate the correct way of doing something, and can create defensiveness. Ultimately, non-reporting may occur if a department tries to beat a previous 'last best' and perceive that some reward is possible (even a non-tangible reward such as recognition)
   • general exhortations are unlikely to be effective because people think the message only applies to others
   • it can involve 'horror', which will bring defensive mechanisms into play
   • it will have a different impact upon different groups; that is, a positive effect on those already acting safely, but have a negative impact on those who are not acting safely (Sell, 1977).

3 In a rapidly evolving business climate, the systems, procedures and practices constantly change and it is too costly to learn from mistakes in this type of work environment.
4 In statistical terms, there is not enough information from which to learn by simply analysing mistakes. Thankfully, the number of fatalities, serious injuries and first aid accident is relatively low and many of them are categorized as freak 'one-off' events. This means that the learning exercise is limited and we are not able to make confident predictions about future occurrences from such weak databases. Figure 1.1 provides one example of an accident ratio study conducted by Tye and Pearson in 1974/75 (cited HSE, 1992) based on an analysis of almost 1,000,000 accidents in British industry. They observed only one fatality, three lost time (three-day) accidents and 50 minor first aid incidents, in contrast to 80 property damage events and 400 non-injury/damage incidents or near misses.
5 Not all incidents are reported; as with disease, prevention is better than cure, but it is obvious that the vital information needed to allow early diagnosis is not at present reaching those ultimately responsible for safety. Top management does already realize this paradox. One senior executive recently commented that, when a fatality occurs, 'we investigate the hell out of it. If the same incident, fortuitously, does not result in any injury it is ignored or a lip-service investigation takes place.'

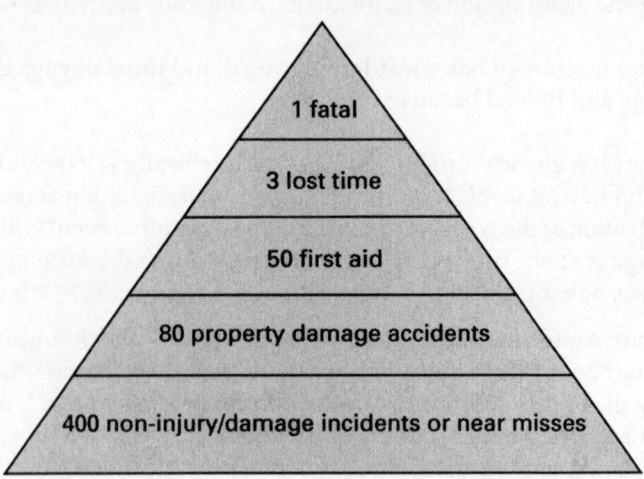

Figure 1.1 *Analysis of almost 1,000,000 accidents in British industry (Tye and Pearson, 1974/5; cited in HSE, 1992)*

This same executive recognizes that what is needed is a 'proactive' rather than 'reactive' approach and has bemoaned the fact that it has not happened. Why? To be proactive you need information before accidents occur. The system has to encourage rather than discourage the reporting of accidents and near misses.

In contrast to the above, the behaviour-based approach to safety improvement is *proactive* and encourages the work group to consider the potential for accident involvement, and their own behaviour as 'safe' versus 'unsafe' *before* someone gets hurt. It develops a practice of investigating 'near miss' or 'dangerous occurrences', by providing a vehicle for reporting and a systematic approach, which takes account of organizational change. This is very important since it is acknowledged that near accidents are excellent predictors of future accidents (Tarrants, 1980). It is our experience that these types of events are not always reported, and are even covered up, because the employee fears blame or punishment. A negative safety culture will continue to exist while the workforce feel unable to report events in a climate of openness and trust. To remove the emphasis from the reporting of accident occurrence as a means to encourage improvement to safety, we concentrate on measuring safety behaviour and the improvement of safety performance through both feedback and goal-setting techniques. Thus, we move from a focus on something negative (nasty and potentially de-motivating) to something positive (pleasant and potentially motivating). The behaviour-based approach also permits us to be specific about desired behaviour, rather than use propaganda that tends to use generalizations.

From a focus on attitudes to a focus on behaviour

The main focus of the traditional safety campaign is on *changing attitudes to safety*. Many organizations employ safety campaigns of one form or another. Often these employ audiovisual methods such as poster campaigns and videos. The intention of these campaigns is two-fold. First, they are an attempt to change people's attitudes towards safety, and secondly, they are an attempt to raise people's awareness of safety issues. Underlying both of these is the assumption that changing people's attitudes, or raising their awareness, will cause them to behave more safely. The experience of many organizations is, that whilst these campaigns may have a short-term effect, the long-term effect is negligible. Three main problems exist with this approach:

1 You cannot see or measure an attitude; and if you can't see it, you can't measure it; if you can't measure it, you can't control it.
2 People often behave in ways that are inconsistent with their expressed attitudes.
3 Raising safety awareness does not always indicate how one should behave differently!

For example, if at the end of a driving test the examiner told you that you had not reached the required standard to be allowed a full driving licence because you had a 'bad attitude to driving', you would be extremely perplexed and annoyed. This is because you would expect an examiner to give you feedback about the specific behaviour that was noted to be of an unacceptable standard; for example, 'not using the rear view mirror before indicating to turn right'. Also, the Department of Transport often use digitalized message systems on motorways, which exhort us to 'Drive Carefully'. However, 'careful' is a relative term and rarely do we set off on a journey with the intention of *not* being careful. Accident statistics would suggest that we do not know what 'drive carefully' means and the notice fails to provide us with this information on how to behave in order to avoid an accident! (The underlying theoretical principles on 'attitudinal' approaches to safety improvement are fully explained in Chapter 2.)

The behaviour-based approach to safety avoids the above pitfalls by placing the *focus on behaviour*. This would seem to be an eminently sound idea since it is suggested that approximately 96 per cent of all accidents, dangerous occurrences and near miss events are attributable in some way to human error or behaviour. For example, only about 2 per cent of road accidents in Great Britain are attributable to the mechanical malfunction of vehicles, the rest are due to errors of judgement and, presumably, to not being 'careful'! So, why not focus the same amount of attention on 'behaviour' as paid to the investigation of occupational accidents and incidents. Figure 1.2 illustrates an amended model of the 'accident iceberg' which incorporates an additional layer to the traditional Heinrich (1951) triangle.

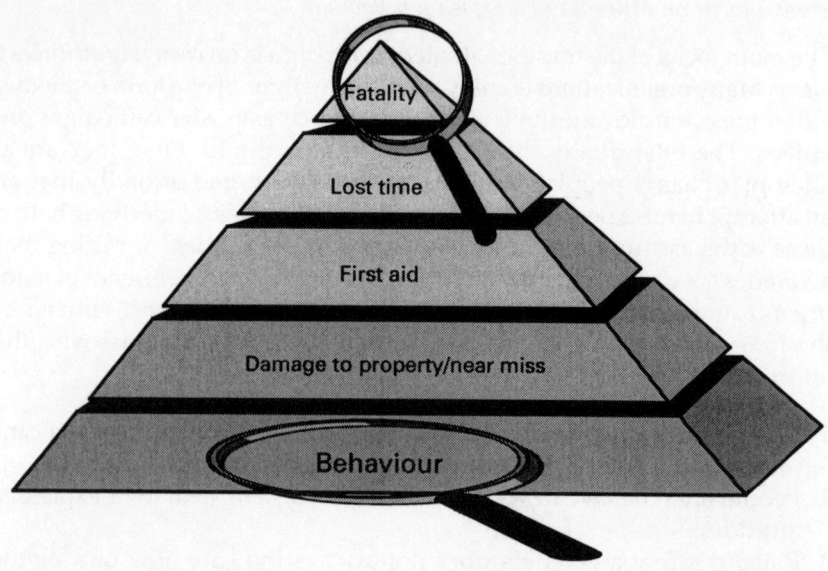

Figure 1.2 *The accident 'iceberg'*

This recommendation is consistent with the guidance provided by the Health and Safety Executive, who state that 'effective health and safety policies will therefore have to examine *all* unsafe events and the behaviours which give rise to them, both as a means of establishing control and as a means of measuring performance' (HSE, 1992). However, we are suggesting that 'behaviour' is measured as either safe or unsafe, before an incident occurs rather than investigating it after the event. Indeed, the organization contracts with, and pays the employee to behave in certain ways. A job description is written in behavioural terms and makes no reference to the attitudes held by the person. Therefore, it is more appropriate and correct for the company to discuss the desired ways of safe working in strictly behavioural terms.

Safety – from the responsibility of the few to the responsibility of everyone

Traditional approaches to safety improvement are typically management driven. However, as one manager said,

> when there is blood on the floor it is not likely to be mine . . . but I still feel very distressed and angry when one of the men is hurt and obviously I want to do all I can to prevent such an incident happening again. Because this state of affairs exists I would prefer everybody to be involved in improving safety and having the opportunity to ensure their own safety and the safety of those working in the immediate vicinity.

The behaviour-based approach to safety improvement leads to an increase in responsibility and ownership of the safety process, at all levels in the organization, because it is *shop-floor driven*, but supported by the management and supervisory structures. As one general manager of a plant employing 520 people stated,

> our traditional approach to safety focused the eyes of my management team and our safety committee on safety issues. Now [after the behavioural approach was introduced], shift by shift I have around 500 pairs of eyes fixed on safety and aware of the links between behaviour and accident occurrence.

Indeed, many companies are trying to empower their workforce as part of a business strategy, often without much success. One of the reasons for this situation is that employees acknowledge that they are empowered, but they do not know what this means in terms of the way in which they are expected to behave on a daily basis. Neither do they believe that they are trusted by management. The introduction of a behaviour-based approach to safety improvement provides a vehicle, system and structure which is driven by employees, for employees. The aim is to improve an aspect of organizational life to which they can relate, namely their own safety! Thus, it empowers the individual by providing 'a feeling of job ownership and commitment, brought about through the ability to make decisions, be responsible, be measured by results and be recognised as a thoughtful, contributing human being' (Sykes, 1996). The approach also helps managers and supervisors to translate empowerment into practical things that they can do since it defines boundaries and roles, and requires them to support and facilitate safety improvement by removing, or helping to remove barriers to safety performance.

As we have already noted, in a climate and culture characterized by change, typified by a streamlined (down-sized!) workforce, it is clear that a high proportion of both managers and supervisors are working in a high demand, high pressure work environment. They are kept busy with ever-increasing paperwork demands and consistently report spending less and less of their time on the shop-floor or at 'the sharp end'. Thus, the amount of time spent inspecting or monitoring safety performance is minimal. They simply are not visible for much of the time and so shop-floor employees must be motivated to take responsibility for their own safety. Quite clearly, a poor safety culture will exist if the workforce believe that the only reason for acting safely is the visible presence of a manager or supervisor. However, many managers do feel some discomfort because they have good reason to believe that safety practices are not adhered to in their absence. The authors can provide some validation of these fears since discussions with shop-floor employees quite often indicate that a much lower level of safety performance exists on the 'night-shift' or at weekends when fewer managers are likely to 'walk' the shop-floor.

The behaviour-based approach provides a structure and system which helps employees to take responsibility for their own safety and be empowered to improve the safety performance of their immediate work group.

A move away from the 'boom and bust' approaches to safety improvement to 'continuous improvement'

Typically, many companies wait for 'something to happen' before they take action. Usually this occurs when a situation reaches an unacceptable level. Figure 1.3 illustrates this concept. When accidents (or production problems, or quality issues) are at a very low level management tends to be relaxed and their attention is focused on some other 'alligator which is trying to sink their canoe'! Essentially, 'their eye is not on the safety ball'. For much of the time accident levels remain at an acceptable rate which is 'comfortable' for management. Thus, it is known as the 'happiness zone', wherein managers adopt a recording and monitoring role. However, as the rate of injury or events rise to an unacceptable level or 'something happens', the attention of management is triggered and this results in some form of action. This behaviour takes various forms, but is normally typified by much shouting, banging fists on tables, and exhortations such as, 'we can't go on like this . . . things have got to improve around here!' Everyone is told they must improve the situation and this is accompanied by a flurry of traditional safety interventions (posters, videos, training initiatives etc.). Therefore, for a while, many more people have their eyes on the safety ball! Thus, injury rates decline, performance improves and everyone feels happy and smug until management relaxes and eventually withdraws attention and moves on to other problems. However, in time, accidents start to happen again and the cycle re-starts. Within this traditional approach, safety improvement is a 'boom or bust' activity because only a few people are charged with the responsibility for safety.

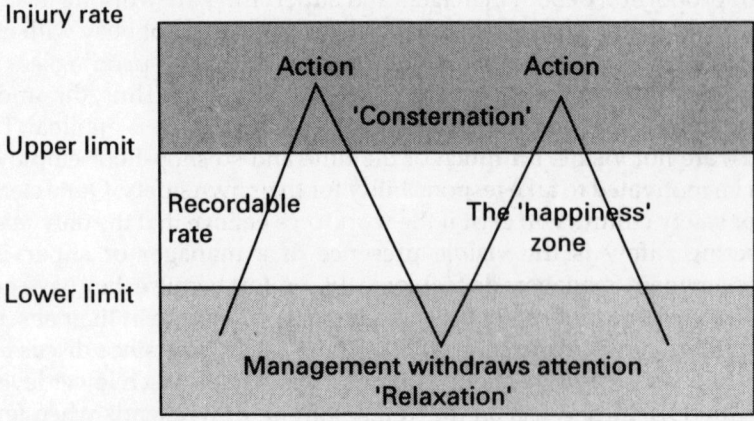

Figure 1.3 *The accident behaviour safety cycle*

In the behaviour-based approach to continuous safety improvement, many more people take responsibility for safety and 'keep their eye on the ball', every day, every shift. Shop-floor personnel take turns to act as observers within their own work group and continually measure safety performance through the use of a checklist which is regularly updated to take into account the changes that might take place in the work environment. In these ways the approach stays alive and is not allowed to become stale or jaded. Neither can it be labelled as a 'flavour of the month' campaign.

The Behavioural Approach: Basic Principles

In summary, it is our belief and experience that most organizations try to improve safety using methods that have only a temporary impact and utilize systems that may actually discourage the very behaviour they are trying to encourage. The behavioural approach represents an alternative way of understanding why people behave as they do, and how they can be influenced to behave in ways that are safer. It can be seen that the behavioural approach to safety differs from 'traditional' approaches in two simple, but important, ways.

First, it concentrates on *behaviour rather than attitudes*. As we have seen, many safety campaigns concentrate upon trying to change people's attitudes, in the hope of thereby influencing their behaviour. The assumption underlying this approach is that attitudes cause behaviour. This assumption is, however, inaccurate. There is considerable evidence showing that attitudes have a tenuous relationship with behaviour. Indeed, they often express how we think we would like to see ourselves behaving rather than how we actually behave. For example, attitude surveys suggest that the general public is against the invasion of the royal family's privacy by the press. However, newspaper editors have firm evidence that the pictures and stories resulting from such invasions do increase sales. Another assumption is that the causal link is from attitudes to behaviour, that is, attitudes cause behaviour. This is, however, only part of the picture. There is evidence from psychological research to suggest that behaviour can influence the formation and change of attitudes. For example, before the wearing of car seat belts was made compulsory, the government spent considerable sums of money trying to change the public's attitude to seat belt wearing. Despite many different campaigns, the level of seat belt usage remained depressingly low. Upon the introduction of legislation making seat belt wearing compulsory, with financial and other penalties for non-compliance, usage leapt to almost 100 per cent. Although we have no direct evidence, we suspect that this change in behaviour also produced a change in attitude toward seat belt usage.

There are a number of possible explanations as to why the way we behave should influence our attitudes, but perhaps the most appealing involves the concept of consistency. We like to be consistent and keep our behaviours and attitudes in line with each other. If we behave in a particular way this must be a representation of our attitudes. These attitudes versus behaviour causation discrepancies explain why traditional approaches to safety have an inherent weakness. The improvement of safety via attitudes not only has to cope with the problems of the tenuous relationship between attitudes and behaviour, but also with the fact that attitudes are notoriously difficult to change. Concentrating on behaviour, on the other hand, misses out this weak link. Accidents are reduced by getting people to behave safely, not by trying to change their attitudes.

The second way in which the behavioural approach is different is its *emphasis on the encouragement of desirable behaviour rather than the punishment of undesirable behaviour*. Organizations may argue that they do indeed encourage desirable behaviour (that is, a good accident record) but, as we suggested previously, this is often counter-productive. The behaviour that needs to be encouraged should be very specific, for example, defining the wearing of hard hats and safety footwear in specific work areas. In many organizations the emphasis of safety systems is often upon the use of discipline and punishment for non-compliance, rather than rewarding compliance. Managers rarely congratulate individuals for wearing their hard hats, but rather punish those who do not. Unfortunately for managers (and perhaps society in general), the ways in which rewards and punishment influence behaviour are not the same, as we will see in Chapter 3.

To summarize, in the behavioural approach the emphasis is on the encouragement of safe behaviour, not the changing of attitudes or the use of discipline.

There is one additional principle that we will be adopting. Whilst it is not a formal part of the this approach, it adds, we believe, to the efficacy of the system to be developed. This principle is that of participation. The reason for its inclusion is two-fold.

First, it is our belief and experience that *the individuals who know most about working safely are those who are actually involved in doing the job.* In doing this we are not trying to deny that 'experts' have an important role to play. For example, it is very clear that manual workers should receive training in how to lift heavy loads correctly. However, if we wish to know whether this training is actually used by those on the 'shop-floor', and if it is not, why it is not, then the people to ask are those actually doing the job.

Secondly, there is clear evidence to show that, if any scheme is to be successful, those who are going to have to carry it out need to be committed to that scheme. This is especially necessary where the scheme is to be self-monitoring. Whilst many organizations make positive noises about participation, we suspect that few actually practise it to any great extent. In our experience, we have uncovered a vast wealth of knowledge concerning best practice, by simply asking people for their perceptions of the safe and unsafe way of doing their jobs. Many organizations using the behaviour-based approach to safety improvement have reported this to have been one of the major benefits of the process. Many significant improvements have been made to both system and practice by this means.

In this chapter we have discussed the rationale for the need for change from a traditional approach to safety management at work to a behaviour-based approach. In the following chapter we discuss in more depth some of the problems involved in using the concepts of attitude and personality as an explanation of behaviour and introduce the notion of risk perception. In Chapter 3 we move on to present the theoretical background to the behavioural approach, before describing its application to safety in Chapter 4.

2

PERSONALITY, ATTITUDES AND CLIMATE

As we have seen in Chapter 1, the vast majority of accidents are caused by people. Sometimes their behaviour is a result of ignorance, but more often it is not. For example, drivers are taught how to drive and have to pass a test, usually having both theoretical and practical elements, before they are allowed to drive unaccompanied. Despite this, road accidents continue to occur. Given that the majority of accidents are caused by knowledgeable people behaving in ways that lead to accidents, how can we understand what caused them to behave as they did?

Attribution Theory

In order to control our physical environment we need to understand what causes the events in which we are interested. Much of science and technology is concerned with seeking answers to questions of causality. What causes hurricanes, or disease, or cycles of economic activity? However, perhaps the major influence on all our lives is the behaviour of other people. The example given above of cycles of economic activity is such a case. It would be difficult to argue that such cycles are caused by predominantly physical factors. The dominant cause appears to be that of 'confidence' – a psychological factor. In other day-to-day situations the ability to predict or control the behaviour of other people is crucial. Much management activity is concerned with such control. The area of psychology that deals with how we decide the cause of people's behaviour is *attribution theory*. It is concerned with understanding how we attribute causality for people's behaviour (including our own).

Attribution theory suggests that there are two main types of explanation for the causes of behaviour. (This is the case whether we are considering safety or any other aspect of behaviour.) The first type of explanation is one that is accepted, almost without question, by the 'man in the street'. It is that a person's behaviour is caused by characteristics such as their personality or their attitudes. Thus accidents are perceived as being caused by people being 'accident-prone', 'foolhardy', 'negligent', or 'having a poor safety attitude'. In attribution theory terms these explanations are referred to as *internal* attributions. Contrast this sort of internal explanation with the second type. This sees behaviour as being caused not by internal, but rather by *external* factors. Because, as we will see, there is a strong tendency for us to prefer explanations of the first type, we will consider a little further the explanation of behaviour as being externally caused.

Let us take an example from the area of road safety. Each winter, almost without fail, there will be at least one multiple crash on Britain's motorway system due to drivers driving too close to the vehicle in front in conditions of poor visibility. The newspapers' coverage of the story is almost as predictable as the crashes – 'Motorway Madness' the headlines will scream. The implication is clear; the crashes were caused by people behaving like madmen – an internal characteristic of the individuals concerned. It is unlikely, however, that *those involved* will subscribe to this description. It is far more likely that they will describe their behaviour as being determined by them reacting to the demands of the situation: 'If I leave a large gap between me and the vehicle in front, another driver will pull into it and I'm back driving too close again.' Or, 'If I leave a gap in front, the driver behind gets annoyed.' These explanations see the behaviour as being caused by external, rather than internal, factors. The person sees themselves as responding to the demands of the situation, rather than their behaviour being driven by internal factors.

Which of these alternative explanations is correct? It could be argued that the explanations offered by the drivers in the crash are defensive in nature. It is more comfortable to blame external factors, for example the situation or other people, than it is to blame oneself. (As one of the present authors has said, 'I can be downstairs washing up whilst my wife is upstairs cleaning. If I break a plate it can sometimes take me as long as 30 seconds to figure out why it was her fault!') However, there is considerable evidence from social psychology and psychological experiments to show that there are a number of systematic biases that influence our attributions, both for our own behaviour and that of others. One such bias has been hinted at above. It is the *self-serving* bias – we tend to blame others for our failures, and take credit ourselves for our successes.

Perhaps the most important bias, however, is in our attribution of the causes of other people's behaviour. There is compelling evidence to show that, when doing so, we *over-estimate the influence of internal factors* on behaviour, and *under-estimate the influence of external factors*. When accidents occur, therefore, the evidence suggests that we concentrate too much on the personality and/or attitudes of the 'culprit', and not enough on the external influences on their behaviour. For example, it is well known that there are accident 'black spots' on the roads. If accidents were caused solely by 'bad' drivers, this would mean all these bad drivers having their accidents on the same short stretch of road – statistically highly unlikely. (It is perhaps interesting to note, in passing, that this tendency varies across cultures. It is most prevalent in the highly 'individualistic' Western countries. It is far less prevalent in the more 'collectivist' cultures of the Far East.)

As well as there being a strong tendency for us to see the behaviour of others as being caused by internal factors, there are other strong pressures that may push us towards accepting such internal explanations. In a safety context, being able to place the blame on the weaknesses of an individual has its attractions. If the individual is to blame, the organization is 'off the hook'. The fear of legal action, either criminal or civil, is largely removed, and the organization need not look too closely at the possible contribution of physical or organizational factors to the accident. Internal explanations, however, are not without their problems. If behaviour is caused by such internal factors as 'accident proneness' or 'poor attitudes' there may be considerable difficulties in achieving any changes. As we will see, such internal factors are notoriously resistant to such change. For these reasons, behavioural approaches concentrate not on internal factors such as personality and attitudes, but on behaviour.

Safety and Personality

In psychological jargon, such personal characteristics as 'accident prone-ness' or 'carelessness' are referred to as personality 'traits'. Examples of such traits are too numerous to mention. Almost any description of a person in terms of their personality will involve the use of a trait. Descriptions such as aggressive, outgoing, uncooperative, friendly etc. are all examples of personality traits. It has been said, not without justification, that the trait is the layman's unit of psychological currency. Unfortunately, if we see behaviour as being caused by traits there is little that we can do to change them. A person's personality is very stable over long periods of time and is very resistant to change. (We are reminded here of the old joke about psychologists: 'How many psychologists does it take to change a light bulb?' 'Only one, but first the light bulb has got to want to change itself!') The old joke, however, contains an element of truth. Changing a person's personality is something that is likely to require expert help, a long time and the commitment of the individual concerned. (It is perhaps worth noting that a description of someone in terms of 'personality' is, in fact, merely a generalization of how they *behave* across a range of situations.)

An interesting example of what appears to be a personality-related safety issue was raised when one of the authors was giving a presentation to a group of safety professionals. A member of the audience initiated discus-sion of the problems associated with having an employee who was an alcoholic. Such an individual poses obvious dangers, both to themselves and others. Surely, it was suggested, here was a trait that had to be changed. The behavioural approach would take a different view – one supported by Alcoholics Anonymous (AA). Members of AA accept that they are alcoholics – they announce it each time they introduce themselves at AA meetings. But they also say how long it has been since their last drink of alcohol. In doing so they are accepting that they will always be alcoholics, but that they can refrain from drinking. What is important is not what the person *is*, but what they *do*.

Describing a problem in terms of the behaviour concerned, rather than personality, also brings certain advantages. First, unlike personality traits, behaviour may be changed, in ways we will consider in the next chapter. In addition, when people feel that they are being criticized for being 'who they are', they will become defensive. To be told that you are 'selfish' is a description of you in terms of your personality, which you cannot, and indeed may not want, to change – hence the defensiveness. To be told, on the other hand, that a particular thing that you *did* made someone feel unloved is different. You can make amends for such behaviour and promise not to repeat it in the future. You cannot promise to change your personality! Most people are not stupid but all of us can, on occasion, *behave* stupidly. Dealing with behaviour, rather than personality, also has fewer ethical complications. If behaviour is caused by personality, then change can only be effected by changing the cause, that is the person's personality. What right does an organization have to try to change someone's personality? It is not their personalities that an organization buys from its employees – it is their behaviour. A job description is written in behavioural terms and the culture and climate of the organization provides the individual with rules and informal instructions about the code of behaviour that is deemed to be acceptable within the company. If the behaviour is appropriate and acceptable, then that is all that matters.

The outlook is rather more promising if we see behaviour as determined by 'attitudes', as they are more susceptible to change. However, even the changing of attitudes is not without its problems.

Safety Attitudes

There are two main problems associated with trying to influence safety behaviour by changing attitudes. The first such problem is that the relationship between people's attitudes and their behaviour is less than perfect. For example, attitudes to political parties at election times are only weakly related to voting behaviour. (In fact, at the 1992 general election in the UK, the exit polls gave a wrong prediction of the result. It was most probably the case that people were not accurately reporting how they had voted, let alone their attitudes to the various political parties.) Another example is our reaction to invasion of privacy of the rich and famous by the media. When public attitudes about such invasions are surveyed the outcome is very clear – between 70 and 80 per cent say they are against such acts. Yet, as was mentioned in the previous chapter, every newspaper editor knows that sales will rise considerably if they print such material!

The second problem is that there is considerable evidence to show that the relationship between attitudes and behaviour is more complex than a simple 'attitudes cause behaviour'. Attitudes and behaviour appear to be inter-related. An example of this complex inter-relationship can be seen in the wearing of seat belts in cars. Before the use of seat belts was made compulsory very few people wore them, despite expensive advertising campaigns. When their use was made compulsory most people complied. It is likely that this has changed people's attitudes to their use. We suspect that, even if the law was repealed tomorrow, most people would continue to wear their seat belts. (It is worth noting, however, that the advertising campaign may have been necessary in order to prepare people for the change. In the USA a law requiring the wearing of crash helmets for motorbike riders was introduced without a prior advertising campaign to convince those affected of its value. The resistance in some states was so intense that the laws were repealed.)

The relationship between attitudes and behaviour, therefore, is not as simple as might initially be thought and hence we will need to examine the subject in a little more detail.

The definition and structure of attitudes

The study of attitudes is one of the most written about topics in social psychology. Questions such as what attitudes are, how they influence behaviour, and how they can be influenced have been the subject of considerable research effort since the 1930s. Originally the term 'attitude' was used to refer to a person's physical posture but it soon came to be used to refer to an internal mental state that is inferred from a person's behaviour. Although there has been considerable disagreement over the years about how attitudes should be defined, most psychologists are now agreed that the fundamental feature of an attitude is that it is an *evaluation*. Eagley and Chaiken (1993) provide the following definition:

> Attitude is the psychological tendency that is expressed by evaluating a particular entity with some degree of favour or disfavour.

In many respects this definition is very close to that of everyday usage. (Recently, however, a new and somewhat different use of the term 'attitude' has being creeping into parlance from the USA. A person is referred to as 'having attitude', normally implying an uncooperative arrogance.) Attitudes may be held about tangible objects, such as football teams, or abstract concepts, such as democracy.

Other people's attitudes are, of course, not directly observable. They can only be inferred from what a person *does* when presented with a stimulus to which they respond. In other words, something stimulates a person's attitude, which then expresses itself in certain ways. These different ways in which the attitude expresses itself have typically been classified by psychologists into cognitive (thought), affective (emotional) and conative (behavioural) responses (Figure 2.1).

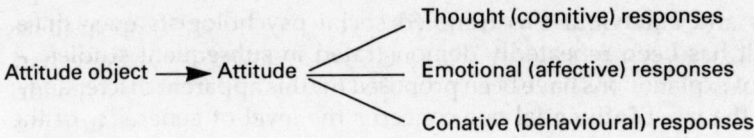

Figure 2.1

Imagine, for example, the reaction of someone who was suddenly confronted by a large snake. If they showed signs of fear and backed away from the snake, these would be expressions of affective (emotional) and conative (behavioural) responses, respectively. In addition, if they were asked what they thought about snakes they might respond that they considered them dangerous, slimy, etc. These would be their cognitive (thought) responses. From these different responses their attitude towards snakes could be inferred. In this case, that their evaluation of snakes was unfavourable. Note that not all three classes of responses are present on each occasion: for example, you might see the affective and conative responses, but the person may not tell you their thoughts about snakes. In addition, their thoughts, or beliefs, about snakes may or may not be accurate; for example, snakes are not slimy.

Attitudes to safety also follow this pattern. Consider, for example, the reactions of an employee, or group of employees, to the introduction of a new safety scheme. It is often not too difficult to infer their attitudes about the scheme from their cognitive, affective and conative reactions to it, that is, what they say, feel and do.

The effects of attitudes on behaviour

Although the definition and structure of attitudes is interesting, the main reason that attitudes are considered important is that they are thought to be a strong, if not the strongest, influence in determining how people behave. Someone with a positive attitude towards safety will, it is assumed, be more likely to behave safely, and more likely to become actively involved in new safety schemes. This, however, is an assumption that has, over the years, been called into question.

Perhaps the earliest example of a discrepancy between attitudes and the corresponding behaviour was demonstrated by Robert LaPiere in a study done in the 1930s. Over a two-year period LaPiere (1934), together with a young Chinese couple, stayed in 66 hotels and other types of accommodation, and ate in 184 restaurants in the USA. At that time there was considerable prejudice against Chinese, yet on only one occasion were they refused service. Following his experiences, LaPiere sent a questionnaire to all the places they had visited, asking if they were prepared to serve Chinese. Of the 50 per cent who responded, approximately 90 per cent said they would not. This demonstration of an apparent mismatch between attitudes and behaviour has troubled social psychologists ever since. Indeed, it has been repeatedly demonstrated in subsequent studies. A number of explanations have been proposed for this apparent discrepancy; perhaps the most influential one concerns the level of generality of the attitude. Attitudes are often elicited towards broad categories of entities, such as attitudes towards physical punishment for crime, for example, corporal or capital punishment. When specific examples arise, however, people's reactions may be somewhat different. The flogging, or execution, of one's fellow countrymen (or, more interestingly, women) in a foreign land often leads to different reactions than might be predicted on the basis of general attitudes towards such punishments. In the LaPiere study, attitudes towards Chinese *in general* may not predict behaviour towards *these particular* Chinese people at *this particular* time and in *these particular* circumstances.

The suggestion that the link between attitudes and behaviour could be strengthened by measuring *specific* attitudes, towards *specific* behaviours was most systematically developed by Fishbein and Ajzen in their Theory of Reasoned Action (see Eagley and Chaiken, 1993). The assumption behind their theory is, as its title suggests, that the decision to behave in a particular way is the result of a deliberate, and fairly logical, consideration of the possible outcomes of that behaviour. This decision making process results in an *intention to behave*. This intention is, according to Fishbein and Ajzen, the best predictor of behaviour. The role of attitudes in the theory is as an influence on a person's intention to behave, which itself will then predict behaviour. Attitudes are not, however, the only influence. The other major influence is that of the person's *subjective norms*. Such norms are the person's standards of behaviour. They are personal, group, or even organizational expectations concerning what is considered to be acceptable behaviour. This can be represented diagrammatically as shown in Figure 2.2. Attitudes and subjective norms are generated in similar ways, according to the theory. An example may perhaps best illustrate the process for each of these factors.

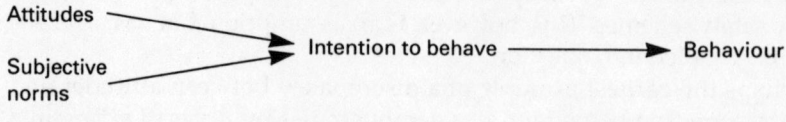

Attitudes

Subjective norms

Intention to behave ⟶ Behaviour

Figure 2.2

ATTITUDES Consider a person's decision whether or not to wear a particular piece of PPE, let us say ear defenders. According to Fishbein and Ajzen, their attitude towards wearing ear defenders will involve:

- beliefs about the consequences of a particular behaviour (in this case the wearing of ear defenders) and,
- evaluations of the consequences of the outcomes of the behaviour.

These two elements are multiplied together. This calculation is repeated for all their relevant beliefs and evaluations about ear defender usage. All these are then balanced against each other to produce the final attitude.

More specifically, let us take examples of each of these elements in turn to see how the process takes place. First, the person has a number of beliefs about ear defenders. For example, they may believe that they are uncomfortable, or not required in the noise levels in which they are working. Secondly, they evaluate the consequences of these beliefs, for example the degree of discomfort caused by wearing them, or the perceived effects on their hearing over a long time period. These two factors – beliefs and the evaluation of the consequences of these beliefs – are then multiplied together. Finally, all these individual assessments are balanced against each other to produce the attitude. For example, the beliefs about the discomfort of wearing ear defenders are balanced against the beliefs about hearing loss if they are not worn.

SUBJECTIVE NORMS The subjective norms are also created by a process of multiplication and summing, but the elements are different. The first element also involves beliefs, but in this case the beliefs concern the reactions of *important* people to the behaviour. For example, how would my workmates react if I were to wear ear defenders? The second element concerns my motivation to comply with the wishes of these other people. For example, do I really care what my workmates think of my wearing ear defenders? Once again, these two are multiplied together and balanced against other, similar, calculations. Other important people might include, for example, partners, close friends, or the safety officer. Their reactions are also assessed and multiplied by the motivation to comply.

These two factors of attitude and subjective norm then influence the intention to behave. This intention is itself a result of a balance between the respective strengths of these two factors. I may, for example, have a positive attitude to wearing ear defenders, yet my intention may be not to wear them because of my perception that my workmates, whose respect I value, would ridicule me if I did so.

The theory has stimulated much research, the vast majority of which has found that the Theory of Reasoned Action strengthened the links between attitudes and behaviour. However, researchers have found other factors, not included in the original theory, that also have an influence. The first modification of the theory was suggested by one of its original formulators, Icek Ajzen. He added the concept of Perceived Behavioural Control (PBC), to create what he referred to as the Theory of Planned Behaviour (see Eagley and Chaiken, 1993).

PERCEIVED BEHAVIOURAL CONTROL As with attitudes and subjective norms, PBC is formed from a number of elements. In this case the elements are:

- the presence or absence of resources and opportunities;

- obstacles and impediments to the behaviour; and
- the perceived power of the person to inhibit or facilitate performance of the behaviour.

In the theory the first two are multiplied by the third to produce the PBC.

PBC, therefore, is an overall assessment of the person's belief as to whether or not the behaviour can be undertaken. This involves perceptions of both the situation and ourselves. The situational variables concern whether the resources are available, together with the nature and strength of any obstacles. The perception of ourselves concerns our ability to carry out the behaviour. All the obstacles to climbing a vertical rock face may have been removed, and all the resources made available to do it, but much will depend upon the potential climber's perception of their own abilities. Diagrammatically, the modified theory becomes as shown in Figure 2.3.

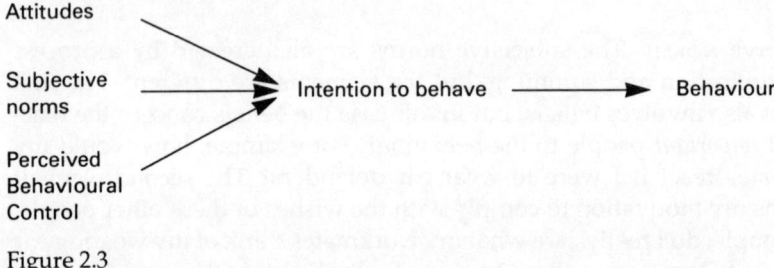

Figure 2.3

Once again, research has demonstrated that the inclusion of PBC strengthens still further the links between attitudes and behaviour. This being said, however, recent overviews of the research on the attitude/behaviour link suggest that, even with the modifications described above, attitudes still only explain between 15 and 20 per cent of people's behaviour. This link is strengthened when the behaviour is novel, that is, we have not encountered a situation before. If this is the case we may reflect on our attitudes before deciding how we ought to behave and the attitude/behaviour link is, therefore, likely, to be strong. Much of our day-to-day behaviour is not novel – we have done it many times before. In such situations the attitude/behaviour link is much weaker – the behaviour is almost automatic, like travelling to work. This suggests that there are some important factors that the theory does not cover. In our view, the two that have particular relevance to safety are the effects of planning, and habits. Indeed, as we will see, habits have a far more powerful influence over routine, automatic, behaviour than do attitudes.

HABITS AND PLANNING Much of safety-related behaviour is, we would argue, habitual. That is, it is not the result of deliberate thought processes, but the result of the regular patterns of past behaviour. Much work is repetitive in nature. Most skills, when once acquired, are controlled more by habit than conscious thought processes. For example, consider your journey to work. Most of the time it is done automatically, only when something unusual occurs is conscious attention brought into play. It is not unusual for people to arrive at their place of work and not remember the journey. Habits, we would argue, have a direct influence on behaviour, independent of attitudes.

The influence of planning on the links between attitudes and behaviour is perhaps not as obvious as that of habits. The effects of planning have been found to strengthen the relationship between intentions to behave and the behaviour itself. Indeed, researchers have found that when plans are strong and precise, they may even eliminate the effects that attitudes have on behaviour. Plans, once developed, have a power of their own.

Putting all these elements together, a composite model is suggested as shown in Figure 2.4. The lines indicate what effects the elements have on other parts of the model. We will return to this model later, when we consider the effects of introducing a behaviourally based safety scheme.

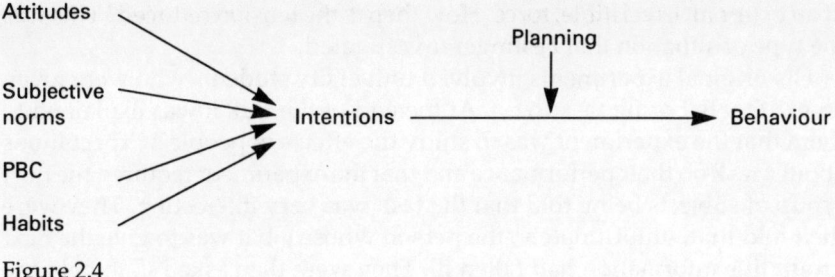

Figure 2.4

The effects of behaviour on attitudes

We have considered the effects of attitudes on behaviour, but what of the effects of behaviour on attitudes? It may appear surprising to consider the effects of behaviour on attitudes, as it is normally assumed that the relationship is the other way around. There is, however, considerable evidence to show that, under certain circumstances, behaviour can have a considerable influence on attitudes. There are two main theories that explain this effect, those of *cognitive dissonance* and *self-perception* theory. For some time it was thought that these theories were in conflict, but recent research has shown that they explain different effects.

COGNITIVE DISSONANCE THEORY The theory of *cognitive dissonance*, developed by Leon Festinger (1962), has been one of the most influential theories in social psychology. Festinger's fundamental assumption, for which there is considerable evidence, is that people like to see themselves, and be seen by others, as being consistent. This also applies to the relationship between attitudes and behaviour. What interested Festinger, therefore, was the situation in which there was a contradiction between a person's attitudes and the way they behaved – in other words, when a person recognized, *cognitively*, that their attitudes and behaviour were in *dissonance* – hence the title of his theory. According to Festinger, a discrepancy between attitudes and behaviour will produce psychological tension. This tension will then produce a psychological drive, the purpose of which will be to reduce, or eliminate, the tension. There are a number of ways in which this may occur.

Consider the situation in which you are being forced to behave in a way that contradicts your attitudes. You may, for example, be faced with the threat of dismissal if you do not behave in a particular way, even though it contradicts your attitudes. In these circumstances the tension is resolved so quickly that it may not even be experienced. The contradiction can be easily justified – you behaved as you did because you were forced to – you had no option. But what about a situation in which you are influenced, but not forced, to behave in a way that contradicts your attitudes? There is no compulsion – you undertake the behaviour voluntarily. In this situation you cannot resolve the contradiction as easily by invoking the justification of an external, irresistible, force. How then is the tension reduced? This was the type of situation that Festinger investigated.

His original experiments involved university students who were given an extremely boring task to do. At the end of the task it was explained to them that the experiment was to study the effects of people's expectations about a task on their performance and that the experiment required the next group of subjects being told that the task was very interesting. They were then told that, unfortunately, the person whose job it was to give the next group this information had fallen ill. They were then asked if, to help the experimenter, they would undertake this job. Nearly all the students agreed. Festinger had thus created a situation in which there would be a dissonance between attitudes and behaviour. The students would tell their fellow students that the task they were about to undertake was very interesting when, in fact, they thought that it was very boring.

There was another element in the experiment. The students who had been recruited to brief their fellow students were paid to do it. Half of the students were paid $20 (quite a large sum at the time of the experiment), the other half only $1. After they had done the briefing their attitudes as to how boring or interesting the original task had been was assessed. What was interesting was the attitude changes in the two groups. It could be argued that the group that had been paid most would have more favourable attitudes towards the task. In general, the more people are paid, the more willing they will be. Thus the highly paid group might be expected to view the boring task more favourably. This, in fact, was not the case. It was the group who had been paid least who saw the task as being most interesting. The theory of cognitive dissonance predicts just this effect! The reason lies in the justification each group can use to explain the difference between their attitudes and behaviour. Consider, first of all, the highly paid group. They can justify the difference between their attitudes and their behaviour as being due to the fact they were being well paid. The low paid group, on the other hand, did not have such a justification available to them. The dissonance could not be resolved by 'blaming' external financial pressures. The only way in which this group could resolve the psychological tension created was by changing their attitudes so as to view the task as more interesting than they originally thought.

The lesson that cognitive dissonance teaches us is seen by some as being counter-intuitive – contrary to common sense. If you want to induce attitude change, get the person to undertake behaviour at variance with their attitude. But in doing so, make as *little* use of external inducements as possible. In this way the person will not be able to justify their behaviour to themselves in terms of the inducements. Their only option is to change their attitudes.

SELF-PERCEPTION THEORY Proposed by Daryl Bem, self-perception theory takes an alternative, and also perhaps counter-intuitive, approach to the nature and change of attitudes (see Eagley and Chaiken, 1993). Bem's theory is based on the way that we infer the attitudes of other people. As we have stated above, it is impossible to observe directly someone else's attitudes. Yet we like to feel that we know other people's attitudes, especially those of people we know well. How do we achieve this? The obvious answer, which was discussed earlier, is that we infer their attitudes from their responses – cognitive, affective and behavioural. Bem takes this process a stage further, and suggests that we infer our own attitudes in much the same way as we infer the attitudes of others – by observing our own responses to the attitude object and then inferring what our attitudes must be! Unlike Festinger, Bem sees attitude change as a system of rational inference, not the result of psychological tensions. These two theories generated much research, often directed at trying to determine which one was correct. Recent studies suggest, however, that both are correct, but that they cover different situations.

It has been known for some time that, for changes caused by cognitive dissonance to take place, the attitude involved had to be important to the person concerned. Often this meant that the attitude concerned had to be important for the person's self-identity. For example, in Festinger's original experiment, most people see honesty as part of their self-identity. Telling lies about the task went against this. This requirement for self-identity to be involved in cognitive dissonance is now recognized as a more general distinction between the two theories. It appears that the important difference is the magnitude of the discrepancy between attitudes and behaviour. Cognitive dissonance works best in those situations where the attitude/ behaviour discrepancy is *large*. Self-perception theory, on the other hand, works best when the attitude/behaviour discrepancy is *small*. We will return to this shortly, when we consider behaviourally based safety schemes.

Changing safety attitudes and behaviour

Let us now turn to safety attitudes in particular, and examine them in light of the above theories. Although we have seen that attitudes influence people's behaviour, we have found that the full picture is more complicated. As well as attitudes influencing behaviour, we have seen that behaviour can influence attitudes. In addition, as shown in the model above, there are other variables that have an influence on behaviour. Let us consider these other variables. As we have suggested above, habits have a strong, direct, influence on behaviour. This effect is likely to be greatest in routine, everyday, safety behaviours, such as the wearing of personal protective equipment (PPE). Norms, especially perhaps the norms of close colleagues, are also likely to have an important influence on behaviour. Norms towards the wearing of PPE, for example, have considerable influence on their usage, especially perhaps with newcomers. If norms can be changed, they may have a dramatic effect on safety behaviour. For example, norms concerning the use of hard-hats on construction sites seem to have changed over recent years. Some years ago they were seen as 'sissy', but their image seems to have changed to being more 'macho', and hence acceptable in the industry. It is thought that norms can be influenced by such factors as the adoption of the changed behaviour by people who are considered 'role models', but the precise causes are not well understood. Not surprisingly, therefore, an easy, and effective, way of directly influencing group norms is not yet available.

According to the model, attitudes, habits and norms influence a person's intention to behave in a particular way. This link is, however, strongly influenced by the person's Perceived Behavioural Control, that is, the extent to which they believe they have the necessary abilities, together with control over the required resources. Individuals will see little point in developing an intention to implement a new safety initiative if they do not believe they have the ability or the resources to carry it through, or if they perceive the obstacles to its implementation as insurmountable. This perception is likely to be based on past experience of failed schemes, and it is often difficult to persuade people that 'things have changed'. This can only be done by demonstrating that resources *are* available and that managerial help will be available to overcome obstacles. This demonstration of managerial commitment will be extremely important at the beginning of the scheme. Absence of management support will only reinforce perceptions that resources are not really available, and that only 'lip service' is being paid to the new scheme.

Once intentions have been developed, these will be likely to be put into action. (Whether or not the intention results in action will be influenced by an assessment of the likely consequences. This, as will be recalled, is an element in attitude formation.) Another influence on the strength of the relationship between intentions and behaviour is the amount and specificity of planning. Detailed planning has been found to strengthen the link.

Indeed, there is evidence that detailed planning can considerably reduce the effects of attitudes on behaviour.

Having considered the structure of safety attitudes, let us consider the lessons from cognitive dissonance and self-perception theories. Cognitive dissonance theory concerns attitudes that are involved in a person's self-identity, and where there is a dissonance between those attitudes and their behaviour. This, for example, may be the case for general attitudes towards personal safety. Many people, we suspect, have positive attitudes towards safety *in general*, yet this may not be reflected in their day-to-day behaviour. This would suggest that there is likely to be dissonance. The theory would suggest that, in order to cope with this dissonance, there would be pressure for the person to change their attitudes. However, this may not take place. If the person sees their behaviour as being caused, not by their attitudes, but by external influences there will be little pressure to change. For example, the dissonance can be justified if the behaviour is seen as the result of organizational pressures, for example pressures for production being stronger than those for safety. (This may be particularly strong in the case of piece-work.) Under these circumstances the apparent contradictions between positive safety attitudes and unsafe behaviour can be tolerated. Cognitive dissonance also shows us, however, that if these external pressures can be seen to be largely removed, the person has no external justification for their unsafe behaviour. The result will be cognitive dissonance, and there will be pressure to reduce it, by changing either their attitudes or their behaviour. The most likely outcome will be that behaviour will change to match their perception of themselves as a 'safe' worker.

Self-perception theory concerns discrepancies between attitudes and behaviour when the attitudes are not particularly strong. It might be suggested that attitudes to specific safety behaviours, such as wearing PPE, fit into this category. I infer my attitude towards wearing PPE from the fact that, often out of habit, I do not wear it. The implication here, of course, is that changing habits may bring about a change in attitudes.

Safety attitudes and safe behaviours

The lessons from the above considerations may now be applied to the introduction of a behavioural system of safety. The behavioural approach, as will be seen in the next chapter, concentrates not on generalities, but on encouraging specific behavioural change. It also places control of the scheme in the hands of those who are most directly affected – those doing the job. This twin-pronged approach has direct, and indirect, effects on the various elements of the model.

Most probably, the primary focus of a behaviourally based approach is that of trying to replace bad habits with good. It does this by specifying precisely the desirable behaviour and introducing systems to try to encourage it. This, of course, increases the level of behavioural planning which, as we have seen, increases the likelihood of the behaviour being undertaken. As mentioned above, management has to remove potential obstacles and make available the resources to implement the new scheme. Observers also need to receive training in the theory and practice of the scheme. All of these increase the level of perceived behavioural control. There is another possible influence of the training course on attitudes. Research has shown that the link between attitudes and behaviour is weakened when people are asked to reflect on the *reasons* for their attitudes. It is thought that this occurs because many attitudes are not the result of deep thought processes, but are based on such things as habits. Asking a person to reflect on the reasons, therefore, may lead them to reassess their attitudes. The training, by presenting information that may not have been considered before, may lead to such a reassessment.

Once changes in specific behaviour have been brought about, the indirect consequences will begin to have an effect. Both cognitive dissonance and self-perception theories tell us that attitudes are likely to change to reflect the safer behaviour. This will especially be the case if management behaviour also changes. Dissonance cannot be blamed on external factors if the management behave in ways consistent with the new approach. By changing the behaviour, the ripple effects will be felt in the other elements of the model. Attitudes are likely to become more positive. In addition group norms are likely to change so as to be consistent with the attitudes of the individual group members.

In summary, therefore, we believe that there are many problems associated with trying to influence safety behaviour *indirectly* i.e. by trying to change safety attitudes. The behavioural approach bypasses these problems to deal *directly* with the behaviour concerned. In addition, however, we have seen above that the relationship between attitudes and behaviour is complicated. Attitudes are not the only factors that influence behaviour. There are also feedback links between the various factors. Changing one of the variables will have an influence, directly or indirectly, on many of the others. Thus, changing behaviour is likely to lead to a corresponding change in such things as habits and attitudes. In addition, our experience suggests that it also leads to a change in safety climate and culture.

Safety Climate and Culture

Although it is not our intention to go into the concepts of safety climate and culture in any depth, it is worth mentioning them as they have received considerable attention in the safety literature. The distinction between the concepts of climate and culture is not clear. Indeed, a review of the two concepts (Rousseau, 1988) found considerable overlap. Despite this, the review concluded that there were sufficient differences between the two concepts for one to be differentiated from the other. Climate, it was suggested, was a descriptive term that applied to the sum of *individual* perceptions of the organization. Culture, on the other hand, was a *group* phenomenon, the expression of strongly held norms, consisting of shared beliefs and values. According to Rousseau, it is possible to have organizations that do not have such strong organizational norms. Thus, whilst all organizations have an organizational climate, not all have an organizational culture. Despite Rousseau's conclusion, in the authors' experience the two concepts appear to be used almost interchangeably, both in research literature and in practice. We propose, therefore, to use the word climate to refer to both.

Safety climate has been defined in slightly differently ways by different workers in the field. However, there does appear to be a general consensus as to its nature. The following definition will suffice as an example:

> the set of values, beliefs, norms, attitudes, and behaviours concerned with minimizing the exposure of employees, visitors, and members of the public to conditions considered dangerous.

Such a definition contains many of the elements discussed above in relation to individual attitudes. In fact, the safety climate of the organization may be considered analogous to the attitudes and personality of an individual. It too represents an inference, based upon the actions of the organization with regard to safety. It is usually measured by surveying the attitudes of individual employees. These are then summated to produce a score for the organization as a whole, which is then taken as a measure of its safety climate. Given these close parallels with individual attitudes, it will come as no surprise to find that the present authors do not concentrate on trying to change safety climate directly. Rather, we believe that concentrating on changing behaviour is more effective.

This does not mean, however, that we see no value in such studies. Indeed, some very well validated attitude survey questionnaires exist, for example the SAQ (Safety Attitude Questionnaire, Donald, 1994). Their value, for us, lies in their use as a diagnostic aid. Many of these attitude surveys contain specific dimensions. The SAQ, for example, has a number of scales assessing such issues as management/supervisor and worker satisfaction with safety systems. Such sub-scales allow a safety profile of the organization to be produced, indicating areas of strength and weakness, which may then be addressed more specifically. Inevitably, however, the method of addressing weaknesses is by changing *behaviour*. We doubt whether climate can, in fact, be changed other than by changing behaviour. Simple exhortations from senior management for a climate change have, in our experience, almost no effect. The value of such measures as the SAQ is in assessing the current level of safety attitudes and indicating which areas need attention. Then, once the suggested intervention has been carried out, they can again be used to see whether the desired effect has been achieved.

Thus far we have considered a number of possible explanations for why people behave in unsafe ways. We have seen the problems of trying to change personalities, attitudes and organizational climate – all of which, it could be argued, have an influence on safety behaviour. There remains another influence on safety behaviour that appears to be common to us all – how we perceive, assess and respond to risk. As we will see, there are common biases in our perception of risk and how we make decisions in risky situations.

Perception of Risk

Almost everything we do in life, whether it be at work or not, carries some degree of risk. Obvious examples include travelling, but accident statistics indicate that even at home we are exposed to significant levels of risk. Most of this we accept as part and parcel of everyday life and hence we do not even reflect on the risk involved. Sometimes, however, we do focus on the risk involved in particular activities before deciding whether or not to undertake them. For example, it is reported that many elderly people will not venture out of their homes for fear of being the victims of violent crime. Similarly, many parents are unwilling to let their children walk to school alone, for fear that they will be abducted, or worse. Given recent health scares the decision as to whether or not to eat certain foodstuffs also becomes an issue involving risk assessment. When people are considering whether or not to undertake behaviours that carry an element of risk, no matter how small, how do they assess the risk involved, and are their assessments accurate?

When assessing the level of risk there are two main factors that people take into account. The first of these concerns the *probability* of something happening, the second, the *cost* if it does happen. These two factors are combined so as to arrive at an overall assessment of the risk involved (some theorists suggest that the two factors are multiplied). There are some interesting differences between 'experts' and the 'layman' in the relative weighting they give to these two factors, probability and cost. *In general*, experts tend to place greater emphasis on the probability of an accident occurring, whilst laymen tend to emphasize the cost of the accident if it does occur. There are many examples of this but two will perhaps suffice. The safety of nuclear power is a much debated issue, especially after the accident at Chernobyl. Public concern about the safety appears to be mostly about the potential costs of an accident, whilst the experts point out the relatively low probability of an accident occurring. Another example is the BSE epidemic in cattle and the chance of contracting CJD the incurable, and fatal, human version. Current knowledge suggests that the probability of contracting CJD from eating beef is extremely low, in the order of several tens of millions to one, yet the effect on beef consumption has been dramatic. The potential costs, it would appear, have more influence on people's behaviour than the probabilities.

This emphasis on probabilities by experts, and costs by lay people, applies to gains as well as losses. One of the authors experienced a possible example of this when he went on an organized outing from his tennis club to a local casino. It soon became obvious that the most popular game for the 'amateurs' of the tennis club was roulette, and that the majority of the bets were on 'long odds', that is, low probabilities of winning but with a high potential win. Watching obvious 'regulars' at the casino, it became apparent that 'expert' gamblers went for such games as blackjack. This game differs from roulette in that it involves an element of skill. It also differs in the probabilities and costs associated with winning or losing. The odds of winning are higher, but the size of the wins lower. Once again, the main influence on the experts is the probability, that on the novices, the potential winnings. The National Lottery is perhaps another example. The amount bet rises considerably in a roll-over week. It is the size of the potential win, rather than the probability of winning, that influences people to buy more tickets.

There is most probably another factor influencing risk assessment. In general, the costs of an accident are relatively easy to assess. The assessment of the probability of the accident occurring is, however, more problematic. As will be seen, biases, caused by the way in which we process information, may mislead us into over-estimating or under-estimating probabilities. These biases are produced by what are known as *heuristics*.

Risks and decision making

As Herbert Simon (1990) pointed out some time ago, human beings are 'limited capacity information processors'. Human beings, like computers, do not have an infinite capacity for information processing. Both computers and human beings are faced, however, with a 'real world' that requires almost infinite information processing capacity to solve even the simplest problems. As Simon points out, the possible permutations after the first pawn has been moved in a game of chess are beyond the processing capacity of super-computers. Yet, humans 'play' chess and computers can 'do' chess. Both manage to do this by making approximations and looking for patterns. In effect, by making a 'best guess' based upon limited information. The systems that produce these guesses are known as heuristics. They are rules, which control how judgements are made and decisions reached. These rules are generally simple but fast, allowing rapid adjustment to changing situations. In general, they serve us well, but their simplicity can be the cause of error under certain circumstances. As an example, consider the perception of the level of violent crime in the UK, an example not unrelated to personal safety.

Crime statistics, especially official police figures, are notoriously unreliable for two main reasons. First, many crimes are not reported, presumably because the victims are not motivated to do so. Secondly, the pressure on the police is to reduce reported crime. Given that this is so, there is a temptation for the police, where possible, to under-report. There are some figures, however, that are less vulnerable to these pressures. Murder, for example, is always reported. In addition, the government finances the British Crime Survey. This operates by interviewing a very carefully chosen representative sample of the population each year to see what, if any, crime they have personally experienced. The figures produced are usually higher than the official police figures, and are generally accepted as being a more accurate reflection of the actual level of crime. Using these figures a familiar pattern appears. People consistently *over*-estimate the levels of murder and violent crime in society. For example, the murder rate in the UK (adjusted for the size of the population) has remained almost constant since the 1930s. Likewise, the number of children murdered by strangers (as is the case with adults, most children are murdered by people who are close to them) has also remained virtually unchanged. (A notable exception to this was the Dunblane massacre.) As well as over-estimating the level of violent crime, there are also common mis-perceptions of the people at risk from violent crime. Contrary to popular belief, women over the age of 65 are the group *least* at risk from violent street crime. In fact, the group most at risk is young men. Why do these mis-perceptions exist?

One explanation, for which there is considerable research support, concentrates on a particular heuristic – *the availability heuristic*. The availability heuristic suggests that people's assessment of the probability (and perhaps importance) of events is influenced by the ease with which other instances of the event can be retrieved from memory. This, in itself, is influenced by the number of times the event has been experienced. In other words, the more often an event is experienced, the higher is the estimate of the probability of it happening again. This is common sense. If we experience an event more frequently, it is more likely to occur again.

The availability heuristic, therefore, provides us with a short cut for estimating probabilities which, for most of the time, serves us well. Our estimates of the possibility of a traffic jam at a particular spot, at a particular time, are likely to be fairly accurate and will help us avoid it. However, there is the possibility of us being misled by an artificially high exposure to depictions of events. With violent crime this occurs through the media. The media tends to concentrate on crimes of violence, especially those involving the vulnerable, older women, for example. In addition, the media is now national. It is only in the latter part of this century that this has been so. Prior to television and improved techniques of communication, printing and distribution within the newspaper industry, most press coverage would be of local events. A murder in Kent would be unlikely to be reported in Scotland. This constant exposure makes our recollection of violent crimes much easier – hence our over-estimation of its prevalence. This effect even applies to individual newspapers. People who read newspapers that report a lot of crime (generally the 'tabloids') estimate its prevalence as higher than those who read newspapers that report less such crime (the 'broadsheets'). (This effect remains, even when such factors as level of education are statistically removed.)

The effect is also clearly visible in people's perceptions of the risk of various forms of travel. Except for multiple accidents, deaths in road traffic accidents are little reported. This is despite the fact that, on average, 10 people die on the roads of the UK each day! Air accidents, on the other hand, receive wide coverage. Many people, however, consider air travel to be more dangerous than travelling by road. In the USA it has been calculated that a person's chances of dying in the crash of a commuter airliner (those feeding into hub airports) is lower than their chance of being killed in an accident involving a vehicle drawn by a horse!

There is one further bias in information processing that is perhaps worth commenting on in the context of safety and concerns our reactions to taking risks. The phenomenon to be described is one of a number of 'biases' in information processing that occur when we are making choices. Originally described by two psychologists, Daniel Kahneman and Amos Tversky (1984), the work has had considerable impact on economic models of choice.

Perhaps the best way of demonstrating the effect is to use a couple of the examples used by Tversky and Kahneman. Let us take a situation that requires a choice. In the first case you are presented with two possible options:

(a) You have an 85 per cent chance to win £1,000; or
(b) You have a certain prize of £800.

Which would you choose? Make this decision before reading further.

The second situation is similar to that above:

You have already won £1,000, but are faced with the following choice:

(c) An 85 per cent chance of losing the whole £1,000; or
(d) A certain loss of £800.

The pattern of responses is clear. In the first case the majority of people go for the certain prize of £800, even though statistically speaking the gamble is a better 'bet' (0.85 versus 0.80). In the second case, however, the majority takes the 85 per cent gamble, rather than the certain loss.

This may appear interesting, if somewhat trivial, but the implications can be significant. Consider the following situation. As a Minister for Health, your medical experts have approached you with some information about a new strain of influenza, which will shortly affect your country. From experience in other countries, they can tell you that 6,000 people will die if nothing is done. You are faced with some choices about what course of action to take. The consequences of each course of action are as follows:

(a) 2,000 people will definitely be saved; or
(b) A one-third probability that 6,000 people will be saved, and a two-thirds probability that no one will be saved.

Which would you choose? Make this decision before reading further.

Now you are faced with another choice of courses of action:

(c) 4,000 people will definitely die; or
(d) A one-third probability that nobody will die, and a two-thirds probability that 6,000 people will die.

This choice experiment has been carried out many times, in many different situations, and the results are very robust. In the first case, choice (a), the definite saving of 2,000 people, is preferred (typically 72 per cent choose this option). In the second case, however, people choose (d), preferring to take the gamble that nobody will die (typically 78 per cent choose this option) against the certainty of 4,000 deaths. The interesting fact, as some may have noticed, is that choices (a) and (c) are *exactly the same* (2,000 of the 6,000 people living is the same as 4,000 dying). The same applies to (b) and (d), they too are exactly the same. Logically, therefore, if (a) is chosen then so should (c). But, in fact, most people switch from choosing certainty in the first case, to taking a gamble in the second. Why should this be so?

The important distinction lies in the way the choice is framed. In the first case the choice is presented as the number of people being saved. In the second case it is presented as the number of people dying. In other words, the first choice, between (a) and (b), is about 'gains'. The choice, between (c) and (d), is expressed as 'losses'. What Tversky and Kahneman found was that there is a strong tendency for individuals to be what they called 'risk averse' for gains, but 'risk taking' for losses. Therefore, whether the decision is presented as a loss or a gain will influence what decision is taken. People prefer to hang on to their gains, but gamble with their losses. (This does not apply, however, to events that are almost impossible.) Another way of putting this is that losses and gains are not equally weighted in decision making. Certain gains are weighted far more heavily than probable losses. This can be easily seen in the balancing of 'production' and 'safety'. Cutting corners and ignoring safety will lead to almost certain increases in production. It takes a much larger, and highly probable, loss to tip the decision in favour of safety.

The theories we have discussed above have a number of implications for traditional approaches to accident prevention and reduction. Whilst lay people concentrate on the costs of any potential accident, experts tend to concentrate on the probability of the accident occurring. It could be argued that, when working at our normal jobs, we are experts and hence concentrate on the probabilities of an accident. Unfortunately, as we have seen, our estimates of the likelihood of an accident may be inaccurate. If we over-estimate the likelihood of an accident this will not matter too much. We will err on the side of caution. There are likely to be problems, however, if we under-estimate the probability of an accident occurring and behave accordingly. This is likely to be the case with minor accidents that occur fairly frequently, but are not likely to receive a great deal of attention or publicity. It is hardly surprising, therefore, that there continue to be a large number of minor injuries. Major accidents, on the other hand, may receive a lot of publicity but, fortunately, they occur fairly infrequently.

Even when we do make an accurate assessment of the probability of an accident occurring, there remains the problem of how much emphasis to give it in our decision making. As we have seen, there are common 'biases' in the way that we reach decisions when balancing potential losses and gains. These will lead us to give far more weight to gains in production, rather than the losses associated with possible accidents. Putting these elements together, it is not surprising that traditional poster campaigns have little effect, as they tend to concentrate on the potential losses, for example, injuries, potential death etc., that can result from accidents, rather than highlighting the probabilities. As we will see in the next chapter, this effect can also be understood in terms of behavioural theory. In addition, the techniques we use help to highlight the occurrence of minor behaviours that may lead to accidents, hence influencing their perceived probability.

In this chapter we have given the theoretical rationale behind our belief that the traditional approach to reducing accidents is ineffective. In the next chapter we discuss the theory underlying the behavioural approach.

3

BEHAVIOUR AND ITS CONSEQUENCES: THE THEORY OF THE BEHAVIOUR-BASED APPROACH

The 'behavioural approach' derives much of its theoretical background from a school of psychology called the 'behaviourists'. This school of psychology takes the view that, as it is impossible to see 'inside a person', psychologists should only concern themselves with 'observable behaviour'. (This, of course, is the basis of the attitude/behaviour distinction made in Chapter 2.) This approach is very similar to using the notion of a 'black box'. The black box approach is widely used in engineering: for example, in the repair of electronic equipment, when testing a complicated circuit board, the engineer ignores what is actually happening in the many microchips which may comprise the board. All he does is to test whether certain 'inputs' are reaching the board, and then ascertains whether certain 'outputs' are being produced. (If the inputs are correct, but the wrong output is being produced, the board will be replaced and the faulty board sent for repair or, more commonly, scrapped.) In the same way, the behaviourist school of psychology does not concern itself with what is happening within the internal functioning of the individual or within the brain, but concentrates instead on the inputs (or stimuli) that it receives, and the outputs (or responses) that these produce. For this reason it is sometimes referred to as stimulus–response, or S–R, theory.

There have been a number of psychologists, from Watson and Thorndike in the early years of the twentieth century to more recent figures, who are all considered as behaviourists and who have all made considerable contributions to the field. However, the psychologist whose name is most closely linked with the 'behavioural school' is the late Professor B.F. Skinner. The techniques developed by the 'behaviourists' are known as 'behaviour modification' (BMod for short), and have been used in many branches of psychology, including educational and clinical psychology. Its uses range from influencing the behaviour of disruptive children to the treatment of phobias (irrational fears), for example, reducing the fear that some people have of flying in aircraft. It has also been used in a variety of organizational contexts. When used in such contexts it is usually referred to as 'organizational behaviour modification' (OBMod). It is with these applications that we will be mainly concerned.

The basic premise of OBMod can be stated very simply – so simply that some may say it is obvious:

Behaviour is determined by its consequences.

Specifically, this means that people will tend to repeat those behaviours that produce 'positive' consequences, and not repeat those that result in either no positive, or 'negative', consequences. In addition, they will also tend not to repeat behaviours that produce no consequences at all. In other words, people learn to behave in ways that produce rewards, and avoid behaving in ways that either produce no rewards or even punishment. As mentioned before, this may appear to be simply applied common sense. Inevitably some managers will already be using some of the techniques of OBMod, but without being aware of it or understanding the theoretical background. By providing a theoretical structure, we enable managers to make more effective use of the techniques.

To summarize and clarify, it is useful to look at classification of the possible consequences of any particular behaviour or action. It is immediately apparent that there are at least four possible options. As a result of behaving in a particular way.

- We *receive* something *nice*.
- Something *nasty* is *taken away*.
- Something *nice* is *taken away*.
- We *receive* something *nasty*.

The first two of these will lead to an increase in the behaviour that preceded them, because the behaviour has been rewarded. To use the correct technical term, they are 'reinforcers', because they reinforce the behaviour concerned. The first, giving something nice, is called *positive reinforcement*; the second is called *negative reinforcement*, because something nasty is taken away. The third and fourth are different forms of *punishment*. They will tend to suppress behaviour. All of these may be neatly summarized in a diagram, as in Figure 3.1. There is one additional outcome, however, that will not fit into the diagram, which is if the behaviour is neither rewarded nor punished. This will also lead to the behaviour not being repeated, and is referred to technically as *extinction*.

Most managers, we find, are fairly happy with positive reinforcement, it makes sense that people will tend to repeat behaviours for which they have received something nice. Punishment also is fairly easily understood, even in its two different forms. As an example, think of imprisonment which combines both the two forms. It can either be seen as receiving something nasty (the prison environment), or the taking away of something nice (freedom and that which accompanies it).

The effects of negative reinforcement, on the other hand, are not quite as clear. An example may help in understanding negative reinforcement. Fire and burglar alarms are based on the principle of negative reinforcement.

Nice   Nasty

	Nice	Nasty
Give	Positive reinforcement	Punishment
Take away	Punishment	Negative reinforcement

Figure 3.1 *Reinforcement square*

True, their primary function is to give a warning, but this does not explain why they have to be so loud. The reason for their high volume is that they are designed to drive the occupiers of the premises (or, in the case of burglar alarms, the intruder) out of the building. The loud noise is so unpleasant that people move to get away from it. As they get outside the building, the unpleasantness stops. Their escape behaviour has been negatively reinforced. (This is also the principle behind ultrasonic cat deterrents for gardens, which are now being marketed.) This is also the reason why it is 'the noisy bearing that gets the oil'. Again, the noise may give some warning as to the state of the bearing, but noisy bearings can often run for a long time. What is more likely is that the noise of the bearing becomes irritating. The oil is applied and the noise stops. The bearing has got its positive reinforcement, and the oiler gets negative reinforcement.

Now we have described the range of possible consequences, let us consider how effective each can be at influencing behaviour. So far, what we have described will appear to many to be 'common sense'. When we consider how effective each of the consequences we have described is at influencing behaviour, however, common sense and psychology part company. We think most people will agree, after considering what we have to say, that psychology is more accurate.

The effectiveness of each of the consequences is largely determined by how frequently it follows, or does not follow, each occurrence of the behaviour. These frequencies are referred to as 'schedules'. There can, therefore, be differing 'schedules' of both reinforcement and punishment.

Reinforcement Schedules

The differing effects of various schedules of reinforcement can be seen if we consider two situations. In the first we are trying to encourage someone (including ourselves) to increase the frequency of a new behaviour or pattern of behaviour. In such a situation, where a new skill is being learnt, it is probably most effective to reinforce their successful attempts every time they occur. This is an example of a schedule of *fixed ratio reinforcement*. In fact, if the behaviour was rewarded every time it occurred it would be referred to as FR1 (fixed ratio, every time). If it were to be rewarded after every twenty times it occurred, it would be FR20.

But let us now look at another, more common, situation for maintaining behaviour that has already been learnt. Consider a schedule where the behaviour is rewarded, let us say, every twenty times it occurred (FR20); for example, pulling a lever to obtain a reward. Assume that you have become familiar with this schedule. If the mechanism were switched off, how quickly would you realize there was no point in pulling the lever any more? We suspect some small multiple of twenty further pulls. Now consider the situation where the rewards occur *on average* every twenty times. The reward could be on the next pull of the lever or many hundreds of pulls later. To determine when this mechanism had been switched off, would take a very long time indeed. This is the principle, of course, of the 'one armed bandit' and the National Lottery. (This also explains why you keep being trapped by the company 'bore'. You have only to reward them by paying attention once every so often and they will continue to pester you.) This is called *variable ratio reinforcement*, and is far more effective in maintaining behaviour, even undesirable behaviour, than reinforcement that occurs every time (fixed ratio reinforcement). In terms of the terminology described above, VR20 is far more effective in maintaining the behaviour than FR20. In variable ratio reinforcement the person being reinforced knows that the behaviour will pay off some time, if only they keep trying. Young children are often masters at this technique. If no one is paying any attention to what they are saying, they will keep repeating it until someone gives in – and someone always does!

A nice example (although not one, perhaps, immediately applicable to the average reader) of the difference between fixed ratio and variable ratio reinforcement is by a scheme to improve the productivity of beaver trappers. One group of beaver trappers in Canadian forests were given $1 bonus for every beaver skin, while another, similar group were given the chance to roll dice each time they brought in a skin. If they rolled two successive odd numbers they got $4. The cost of each of the schemes was the same. Productivity in the first group rose by 50 per cent, in the latter group by 108 per cent.

(As well as fixed and variable 'ratios', there can also be fixed and variable 'times'. The effects are similar – fixed time reinforcement is a relatively ineffective way of maintaining behaviour. It remains, of course, the most common way of paying salaries, which is why regular salaries are not an effective way of influencing behaviour.)

Punishment

So far we have considered how behaviour can be encouraged through reinforcement. Let us now turn our attention to how to stop undesirable

behaviour. As we have seen, the strategies available are those of punishment and non-reward. It is important to note that non-reward is *not*
the same as ignoring. Non-reward means that the person gets no benefit
whatsoever as a result of their behaviour. (Ignoring, on the other hand,
implies that someone is receiving a reward for undesirable behaviour
and getting away with it.) All the evidence suggests that non-reward leads
to the behaviour being 'extinguished', whilst punishment merely sup-
presses it. This is not to say that punishment is never effective. Punishment,
by itself, can be effective under certain conditions. (When reading further
you might find it interesting to think of the ways that society tries to
combat crime and/or the ways some organizations try to enforce safety
regulations.)

There are two conditions that have to be satisfied if punishment is to be
effective – the behaviour is punished *immediately* after it occurs, and the
behaviour is punished *every time* it occurs. In other words, if you want to use
punishment as a way of influencing behaviour, you have to be in a position
to catch *every* infringement and punish it *immediately*. If you wanted to use
punishment, by itself, to get people to wear ear defenders, for example, you
would have to have a manager watching every worker all the time they
were at work and be ready to issue some form of punishment, even just a
rebuke, every time they failed to put on their ear defenders in a noisy
environment. Contrast this with the much easier requirement for keeping
the behaviour (even undesirable behaviour) going, for example, variable
ratio reinforcement!

A real example of the differential effects of rewards and punishment
was discovered in a factory in which the authors were implementing a
behavioural safety scheme. Before starting the intervention the accidents
records were examined. A number of occurrences of burns to wrists and
forearms were identified – all in one department. Before knowing the exact
cause, the theory allowed us to predict, in general, what we would find.
Since the behaviour was reoccurring, the punishment of the burn could not
be happening every time. In addition, since the behaviour continued to
occur, there must be a reinforcer keeping it going. The exact cause was later
identified. The final stage of a process to make a thin film entailed the film
passing through a heated roller, onto the final take-up roll. Sometimes the
film broke. If the operator could catch the film before it disappeared through
the heated roller, they saved themselves an hour's work re-threading the
machine. Most of the time they did catch it but, just occasionally, they got
a burn off the heated roller. Variable ratio reinforcement took precedence
over intermittent punishment. In our experience many accidents follow this
pattern. A shortcut is taken because, most of the time it pays off and rarely
leads to punishment. For example, driving too close to the car in front rarely
results in an accident, the payoff being preventing another driver cutting
in and taking the space between you and the other car.

The major point that comes out of the behavioural approach is very
simple.

Because of the ways in which punishment and reward operate, it is usually much more effective to try to encourage the desired behaviour than it is to punish the undesirable!

This does not mean that punishment is *never* effective, only that it is often very difficult to achieve the conditions required for it to be so. There are many examples in everyday life which illustrate this difficulty. For instance, hangovers rarely stop people drinking to excess. The punishment is delayed (it occurs the following morning) and it may not happen at all.

The difference in the way in which rewards and punishments work is known as *reward/punishment asymmetry*. To be most effective rewards should occur on a variable schedule. Punishment, on the other hand, is only effective if administered every time. (This asymmetry appears to have a physical basis. Recent research has shown that rewards and punishments affect different parts of the brain.)

Timing of Consequences

There is another factor that affects the effectiveness of both rewards and punishments. This concerns the delay between the behaviour and the consequences that follow from it. There is a very simple general principle: that is, that more immediate consequences will have more influence over behaviour than consequences that are delayed. As we will see, this has important implications for those trying to improve safety.

Consider, first of the all, the situation in which the rewards are immediate but the punishment delayed. Many health related behaviours such as smoking, over-eating and drinking too much alcohol, fit into this category. The pleasures are immediate, the negative consequences on the other hand are delayed. Unless the person is very strong-willed, the immediate consequences will exert most influence. This influence is also behind 'buy now, pay later' marketing techniques. The pleasure of the product is immediate, the pain of having to pay comes later! The alternative healthy lifestyle, that we all know we should adopt, are examples where the punishments are immediate but the rewards delayed. Taking exercise for those who do not like it is an immediate punishment. The reward of a fitter body is delayed. Again, without a strong will, the influence of immediate punishment will prevail.

This differential effect of the timing of the consequences can be seen in such areas as personal protective equipment. Wearing ear defenders, for example, is initially uncomfortable (immediate punishment). The delayed punishment (hearing loss some 20+ years later) has relatively little influence over present behaviour. The solution to this problem lies in arranging reinforcers for the behaviour that occur more immediately. (This is the technique used by slimming clubs who use social reinforcement on a weekly basis for small, but progressive, weight loss.) The timing effect can also be used to understand managerial actions. Often, the shop-floor's perception of management is that they are more interested in output than safety. This, in fact, often accurately represents the pressures on management. The rewards for immediate production are obvious. The rewards, if any, for a good safety record are usually delayed.

This differential effect of rewards and punishments can be observed every day on the nation's roads. The use of sanctions to influence driving behaviour is often doomed to failure. The rewards for speeding, or going through traffic lights on 'deep amber' are obvious – time is saved. The chances of being caught by the police are minimal. Automatic speed and traffic cameras, however, often have a dramatic influence on behaviour. The reason they do so is because the probability of detection and punishment is high. It is the certainty of detection, rather than the severity of the punishment, that has most influence on behaviour. Two examples from car crime, each with public safety implications, demonstrate this fact.

Drink driving is normally punished by an automatic ban from driving of a year. Although the general level of drink driving has declined over the years there have been occasional increases at certain times of the year. The pressure when this occurs is often a demand for an increase in the length of the ban. Chief Constables realize that this is not what is required. Their traffic committees have said that the most effective way of reducing drink driving is to ensure that the culprits are caught every time. The second example is taken from a Yorkshire Television documentary which filmed the effects of the IRA's attempts to stop 'joy-riding' in Republican areas of Northern Ireland. The sanctions were clear and specific: someone caught joy-riding was given a warning – if they were caught a second time they would be knee-capped, and if they persisted after this they would be killed. The programme interviewed adolescents attending the funeral of one of their friends who had met such a fate. They included one person who had been knee-capped and, because of infection, had lost his leg below the knee. Following the funeral their intentions were to go joy-riding! In the absence of high detection rates, even such draconian punishments were ineffective.

Some Applications of Reinforcement Theory

To show further uses of the theory and provide more insight we will now consider some real-life examples where the use of encouragement might be effectively substituted for punishment. Dealing with lateness and absenteeism provides one such example. A common reaction from management is for each individual lateness, or absence, to be noted but not commented on. Action is left until an unacceptable number of incidents have accumulated. There then, usually, follows a series of stages consisting of a very gradual escalation, from verbal to written warnings, to formal dismissal procedures. Under these circumstances, punishment is unlikely to be effective. For it to be so, the individual should be made aware that each and every lateness will be questioned as soon as it occurs, and that moderate sanctions will be applied. (Very harsh sanctions would most probably lead to avoidance and create further absenteeism.) Other approaches to absenteeism see it as a symptom of some underlying problem. The assumption is that absenteeism is a symptom of dissatisfaction with some aspect, or aspects, of the job. The solution, therefore, is to improve job satisfaction. This may, indeed, have some impact, but there is often a limit to what can be done in improving satisfaction. Shift work still needs to be worked, and many other 'dirty', and undesirable, jobs need to be done. In addition, the rewards for staying away from work are powerful, and rarely capable of being influenced by management.

OBMod, on the other hand, looks not at the *influences* that are thought to underlie absenteeism, but rather at the *consequences* to the employee of attendance or non-attendance. It adopts a 'direct action' model. Let us consider some possibilities. One such example was reported from a factory in Liverpool. The factory was to be closed and the production lines transferred to another part of the country. The workers were under notice of redundancy, but it was essential that production be maintained until the new factory was in production. Unfortunately, the factory was suffering absenteeism levels of 30 per cent and above, due to a 'mystery virus' that appeared to strike mainly on Mondays and Fridays. Because of employment law relating to redundancy, pay could not be stopped for these absences. In order to improve attendance, management instituted a weekly prize draw of £500. Participation in the draw was by means of tickets. Each employee received a draw ticket on each occasion they attended for work on time. Absenteeism dropped to very low levels and the management reported that workers were even turning up on their days off in order to collect tickets. Other schemes have used a cash bonus, paid to every employee who had attended on a number of randomly selected days during a set period – the random choice of days provided the variable ratio.

This demonstrates one of the conditions under which such reinforcement works best – the expenditure of little additional effort on the part of the individual, together with the potential for a large payoff. This is the principle on which football pools and premium bonds work. Indeed, the use of prize draws is widely used as a marketing technique. There are some managers, incidentally, who object to such schemes on the basis that they involve paying people extra to do what they are already paid to do. This is, of course, a legitimate position to take, but in situations where all else has failed adherence to principles such as these may have a cost – in this example, continued high levels of absenteeism.

A good example of how the removal of reinforcers can lead to the cessation of undesirable behaviour was reported by Merseyside police. Previously they had concentrated upon catching those who were stealing car stereo equipment, with little effect. They then switched their strategy to that of identifying those cars whose owners might have purchased stolen stereos. They examined parked cars looking for incongruities – for example, an old car with a highly priced modern stereo – and then contacted the owners for an explanation. As soon as it became known that this was happening, the market for stolen stereos declined sharply. The rewards for stealing the equipment were therefore removed and thefts declined markedly. Unfortunately, however, the thieves then found markets outside the Merseyside area, in nearby cities such as Manchester.

One final example concerns the problems associated with routine maintenance procedures. All the rewards and punishments are geared to encourage short-cuts; if a part is not checked, it will most probably be alright anyway, and the mechanic saves time right now. In addition, there is no reinforcer to encourage staff to carry out the checks as specified. If something does eventually go wrong, then what evidence is there to refute the claim that 'it seemed alright when checked'? The evidence of successive Consumer Association reports on the quality of car servicing by garages lends strong support to our analysis.

The Application of Behavioural Techniques to Safety

The following provides an excellent example of how a change of emphasis, from punishment to encouragement, can have dramatic effects. Two researchers, Dov Zohar and Nahum Fussfeld (1981), implemented a behavioural safety scheme in an Israeli textile factory. Despite very high noise levels in the weaving sheds (averaging 106 dBA), only about 30 per cent of the workers in those areas were wearing the ear defenders that had been supplied. This figure proved resistant to attempts to improve it. Methods tried included those designed to change attitudes and the use of discipline. When asked as to why the ear defenders were not used workers commented that they were, at least initially, uncomfortable. Those who wore the defenders on a regular basis, however, reported that the initial discomfort wore off, and that if they then had to remove them, they became aware of the magnitude of the noise. Many of this minority would now not enter the weaving shed without their ear defenders.

The method that was adopted was to reward the wearing of ear defenders with 'tokens'. These tokens could be saved, and exchanged for goods that were displayed in cabinets in the weaving sheds. The workforce were advised of the nature of the scheme, its start date and the date at which it would finish. In all, the scheme lasted two months. It was also made clear that there was no compulsion to take part, and that non-wearers would not be subject to sanctions. In order to make the reinforcement schedule variable, token distribution was undertaken three times a week on a 'quasi-random' basis. In order to overcome the initial discomfort of wearing the defenders, time bands of increasing length were used during which token distribution would take place. For the first three days distribution was during the first two hours of each shift; during the next three days distribution was during the first four hours, and so on. Within these time bands the distribution was on a totally random basis.

The results of the technique were impressive. Within the first week ear defender usage rose to effectively 100 per cent and remained there for the whole of the two-month period during which tokens were distributed. Even when the token distribution ceased, the level of usage remained at the same level, and did so on random follow-up visits almost a year later. Like the seat belt usage referred to earlier, the wearing of the defenders had become 'self-reinforcing'. People now realized how high the noise levels were when they removed the defenders (they would, of course, be negatively reinforced when they put them back on – the unpleasantness of the noise stops).

We often use the example given above to demonstrate the potential effectiveness of behavioural techniques. However, the example is not typical of many interventions in two respects. First, there is money available to act as reinforcement. This is often not the case, and is an issue we will return to later. Secondly, the desired behaviour is specific and precise and it is hence very clear what behaviour is to be reinforced. This may not always be the case. Indeed, organizations may, for the very best of intentions, reinforce the very behaviour that they wish to *discourage*. Examples may be found in many organizations, but we have taken a particular interest in safety on the oil platforms in the North Sea.

Perhaps the most comprehensive study of safety in the North Sea, and in particular the events surrounding the destruction of the Piper Alpha rig, was provided by the Cullen report (1990). This provided examples, if any more were needed, of how organizations ignore human factors to their, and often the public's, cost. Often, despite the best of intentions, organizations devise and install systems whose effect is to encourage the very behaviour that they seek to discourage. This appears to be what has happened, and is still happening, with safety practices in the North Sea oil industry, where accidents continue to happen despite the millions of pounds that oil companies have spent trying to improve their record, and despite the expressed desire of top management to take steps to improve it.

A common example is the allocation of annual budgets within organizations. Most organizations urge all their departments to be as efficient as possible and to save money. If, at the end of the financial year, an efficient department has heeded this advice and is under-spent on its budget, what happens? In our experience the surplus is often 'clawed back' by the financial administrators and the following year's budget is cut. The department has been punished for doing precisely what was asked of it! A more subtle mistake occurs when organizations reward the achievement of 'final goals' without taking into account the behaviour required to reach that goal. In doing so they sometimes inadvertently reward behaviour that is incompatible with the final goal. What needs to be rewarded is the behaviour required to achieve specific sub-goals. As this distinction is rather difficult to understand in the abstract, let us illustrate with an example which, we believe, has parallels with what is happening in the oil industry.

Some time ago the World Health Organization undertook a concerted campaign to eradicate smallpox. Most cases of the disease occurred in the

'Third World', and the 'front line troops' for the campaign were health visitors. Each of these had a geographical area for which they were responsible. In order to motivate the health visitors, a bonus scheme was introduced. Arguing that the final goal was the eradication of smallpox, a scheme was devised whereby each visitor was rewarded according to the absence of smallpox in their area. However, although the visitors consistently earned good bonuses, smallpox remained endemic. When considered from the visitors' perspective, the reasons for this apparently paradoxical situation become clear. If you are rewarded for the lack of cases, the incentive is to turn a blind eye. When in doubt, don't report. The system is obviously open to abuse. Management finally realized the potential for abuse, and the reward system was turned on its head. Instead of being rewarded for the *absence* of cases, visitors were now rewarded for *finding* cases. The results were dramatic, undiscovered cases now came to the attention of the authorities and could therefore be treated. As we are all aware, smallpox is now officially 'dead'. However, we doubt if this would have been the case had the original reward system not been changed. We believe that the same sort of well-intentioned error is happening on British oil rigs and platforms in the North Sea (as well as elsewhere in industry). From one end of the chain of command to the other, all the rewards are for turning a blind eye to accidents or near accidents. This tone is set at the top, despite the best of intentions to the contrary.

The Department of Energy, quite rightly, puts pressure on oil companies to reduce accidents. The Department's emphasis is upon 'accident-free' operations, usually measured as 'lost-time accident-free periods'. As we will see, whilst this is laudable as an ultimate goal, it is not where the emphasis and rewards should be placed. This method of assessing safety is transmitted down, through the companies' chains of command, to the rigs, platforms and other installations. The goal on the rig, therefore, is to report as few accidents as possible. The pressures on the workers are to turn a blind eye to minor accidents, and especially near accidents. Even the reporting of fairly serious injuries may be discouraged. This pressure is intensified by the rewards for such 'accident-free periods'. For example, drilling crews work as an eight- or nine-man team and rewards for 'accident-free' periods are given to the team as a whole. Following a minor incident (or a potentially major incident that fortunately came to nothing), all the pressures from other members of the team are to not report it. If it is reported the whole team will lose. The pressure is not only the prospect of losing an individual reward, but the guilt of knowing that your fellow workers will also suffer. In some organizations these guilt feelings are enhanced still further, again with the best of intentions. If the team achieves a certain accident free record a sum of money is donated to a charity of their choice. Thus insisting on reporting a minor accident will make starving or sick children the losers!

For sub-contractors the pressures are even more immediate. Their 'accident-free' record is an important factor in deciding whether they will get any further work. In addition, contractors are only paid whilst they are on the rig. If they are sent ashore, their pay stops. An analysis of 'time off' reveals that contractors return to work far quicker than salaried staff following similar injuries. Whilst it could be argued that the salaried staff are making the most of their injuries, we suspect that, because of the pressures of money, some contractors may be returning to the rig before they are fully fit. This may pose an additional hazard to both themselves and others on the rig.

What needs to happen, therefore, is that rewards need to be directed towards the *reporting* of accidents and near accidents, not the attainment of 'accident-free' periods. It is the analysis of information from such reports that will allow preventive action to be taken. Like the health visitors and the eradication of smallpox, if you are to prevent accidents you need information about their occurrence or near occurrence. Not only should workers be rewarded for information about accidents and near accidents, but individuals and teams should be encouraged to produce suggestions for improving safety. The workers at the sharp end have most probably more information and experience than anyone else. At present they are actively discouraged from using that information. Contractors also should not be judged on the lack of accidents to their workers, but for having systems that can identify, and respond to, potentially dangerous situations. Whilst they should retain their punitive sanctions for flagrant breaches of regulations, the Inspectorate should likewise begin to reward companies who report accidents and near accidents. The Norwegian authorities and companies are already attempting to do this. In another context, the air traffic authorities have established a 'confidential hot line' for airline pilots to report 'near misses' without fear of reprisal. The oil companies have set up a similar scheme so that workers can report safety-related information. Unfortunately, this system also punishes the desired behaviour. Calls 'ship to shore' can be overheard and workers are afraid that they will be punished for such 'leaks'. Even if it has never happened, there is a strong perception amongst the crews that if you are caught complaining you will never work offshore again: your card will be stamped 'not required back (NRB)'. If workers phone whilst on shore there is still the possibility of discovery because the source may be traced through the specialist information revealed about a particular rig. In addition, there is no reward for such reporting.

All these examples illustrate the fundamental flaw in the system. Reporting is punished, silence is rewarded. As Peters and Waterman (1982) commented in their book *In Search of Excellence*, the complete absence of errors or negative feedback is a warning symptom of ineffective climates. If the companies wish to become 'proactive', the whole system of rewards needs standing on its head: reward the giving of accident-related information and suggestions, not 'blind eyes' and 'cover-ups'. To reiterate, it is our belief that most companies are well intentioned as regards safety. Unfortunately, these good intentions are thwarted by the very systems that are intended to help their achievement. This problem is not unknown to the authors, even when behaviourally-based systems, concentrating on the reward of desired behaviours, are introduced. We have sometimes been contacted by companies who have recently introduced the scheme, worried that near misses have increased. (This is often not helped by the fact that some companies count near misses as accidents.) As we point out, an increase in near miss reports is a good, not a bad, indicator. It means that people now believe that there is something to be gained in reporting them – action may, at last, be taken to improve safety conditions.

Practical reinforcement

Having considered the problems surrounding the identification of the correct behaviour, let us now turn to the issue of reinforcers. That the use of tangible reinforcers works is, perhaps, not surprising, but what if financial rewards are not possible? What else can be used as reinforcement? There are three types of reinforcers: *primary*, *secondary* and *generalized*. Primary reinforcers are those that do not have to be learnt and are basic to our very existence, such as food and water. Secondary reinforcers, on the other hand, are those that become reinforcers by being regularly associated with the primary reinforcers. In order to survive the infant has to depend upon other people. Through this association other significant people, notably the parents or parent substitutes, become reinforcers. As we will see, such 'social' reinforcers are most probably the most powerful of all. Finally, there are generalized reinforcers, of which money is the most obvious example. Money attains its power as a reinforcer because it can be exchanged for those goods or services that the recipient finds inherently reinforcing.

There is a problem of satiation with primary reinforcers. Once you have had enough food you do not want any more, at least in the short-term. For this reason, generalized reinforcers tend to be the most powerful. (In an organizational context secondary will be most often used.) Social reinforcers, for example attention and praise, are powerful, can be easily administered and do not disrupt on-going behaviour. They are, in fact, 'naturally occurring' in everyday life. Many reinforcers, therefore, have acquired their value as a reinforcer by association with other reinforcers, often early in our lives. Two of the most powerful are praise and recognition, and feedback and the attainment of self-set goals.

Praise and recognition

Within the work situation in general, praise is used far less effectively than it might be. Variable ratio reinforcement suggests that praise should not be used on every occasion, indeed, if over-used it will eventually lose some of its value. This being said, the evidence is that it is hardly used at all. Why is this? A survey amongst managers suggested a number of reasons. Some took the view that it was inappropriate – their employees were already being well paid for what they did. For the majority, however, it was the embarrassment of giving praise; it is somehow 'not macho'. Many, however, thought that they already did give praise. Over 80 per cent of managers in one organization claimed they praised their employees for work well done. The researchers then asked those employees about the praise. Only 14 per cent said their manager actually gave praise. Another reason for not giving praise is that we tend to 'manage by exception'. We shout and complain when things go wrong (the exception), but do we praise when things are going routinely well? You might compare the number of times you yourself have complained, for example in shops, to the council, or to other managers, with the number of times you have written complimenting them when they have been particularly helpful. Praise is a very powerful reinforcer, especially when used on a variable schedule. This schedule should not be random (i.e. the occasional praise for 'work in general'), but should be specific, and very clear to the recipient which part of their behaviour is being praised.

An example of the power of praise as a reinforcer came to the attention of two of the authors on a management training course they were running. The course was for managers working for a large 'catalogue' store. Of these managers, a small number worked in the department responsible for deciding what goods would be included in the catalogue. Each season the contents of the catalogue would be finalized and the head of the department would then attend the next board meeting to present the new season's catalogue. On the occasion in question a new head of department had recently taken over. When his staff presented him with the new catalogue he informed them that, rather than attend the board meeting on his own, they were to accompany him. Having presented the new catalogue to the board, and received their congratulations for its quality, he ushered his staff into the meeting and asked the board to thank them personally. The obvious beneficial effect that this praise had on the staff was still evident over six months later! Praise, properly used, is a powerful reinforcer – at little cost!

There are other reinforcers, also with little cost, that can also be used effectively to influence behaviour. Two of these are 'goal setting' and 'feedback'.

Goals and Feedback

As well as being based on behaviour modification, the behavioural approach to safety also involves aspects of goal-setting theory. Indeed, some theorists would suggest that it actually forms a part of BMod theory. In any case, this theory is one of the most influential theories of motivation. It also is based upon fundamentally simple assumptions. The name most commonly associated with goal setting is that of Locke. According to Locke and Latham (1991), reinforcement comes from the achieving of specific goals. In addition, the harder these goals, the greater the effort and subsequent performance and satisfaction. There is considerable evidence to support these contentions, but first, we should look at how goal-setting works. It does so by:

- directing attention
- facilitating a search for appropriate methods
- mobilizing and maintaining effort.

Directing attention towards the task involved is, of course, necessary for the task to be undertaken. It also has the effect of removing uncertainty as to what precisely is expected. Once attention is directed, there needs to be a search for appropriate methods of task attainment. When the task is clear and precise it is possible to rule out a large number of alternative strategies and concentrate on those most directly relevant. In fact, the more specifically the goal is stated, the more positive the effects. Goals along the lines of 'do your best' are not effective in directing attention and choosing appropriate strategies. They need to be as specific as it is possible to make them, preferably couched in behavioural terms. Finally, goal-setting mobilizes and maintains effort.

Locke originally made a distinction between the different way goals are set: that is, between *assigned* versus *participative* goal-setting. An assigned goal is one in which there is no discussion with the individual concerned. A participatively set goal, on the other hand, involves the individual in decisions about the goal. Surprisingly perhaps, initial studies suggested that whether the goal was assigned or set participatively had little effect on performance and goal attainment. Further research suggested that this is indeed the case, as long as the goal is *accepted* by the individual to whom it applies. This has led to considerable discussion and experimentation about the factors that influence goal acceptance.

One explanation was provided by Locke and his co-workers. For Locke, perhaps the central concept of goal-setting theory is that of 'goal difficulty'. Participation, it is argued, leads to an increase in the level of goal difficulty, thus leading to improved performance. As long as assigned goals are accepted, more difficult goals will also lead to improved performance. Hence there is no difference in the effects of goals that are participatively set, and assigned goals that are accepted. Locke and his colleagues have shown that this does appear to be the case. For similar levels of accepted goals, there was no difference in performance between those that had been assigned and those that had been set participatively. Other researchers have found, however, that those who have been involved in setting their own goals perform better. According to this view, participation affects performance not through the effect of more difficult goals, but through the *commitment* generated by the participation. It is generally accepted that participation improves commitment. Many people will have experienced this effect. The opportunity to express their views and to have some control over the goals they are set seems to have a positive effect on most people's commitment. Recently, this debate has been resolved. It appears that the major difference between assigned and participatively set goals is the extent to which a reason, or rationale, is provided for the goals being set. It would appear that the critical feature required for goal acceptance is the provision of a *reason*.

So far, nothing has been said about letting people know how close they are to achieving their goal – in other words, *feedback*. It has been shown that feedback by itself can lead to higher levels of performance. Consider, for example, children (and, perhaps, adults) playing with computer games. Often these games do not involve any other person, and there is no tangible reward, for example, money. What then is the reinforcement that keeps the behaviour going? A reinforcer so powerful that it has led some experts to suggest that the games can become 'addictive'? The main reinforcement appears to be immediate feedback about performance. That feedback alone can influence behaviour has been shown in many organizational contexts. In safety as well there is evidence for its effectiveness. For example, it has been demonstrated that giving drivers feedback that they are exceeding the speed limit reduces average speeds. Such systems have now been introduced on some UK roads. Both goal-setting and feedback can be powerful reinforcers; the two together, however, are more powerful still.

We have now described the elements of the behavioural approach to safety. It can be summarized as:

> The reinforcement, by the use of praise, goal-setting and feedback, of precisely defined behaviours that have been identified as contributing to safe working practices.

In the next chapter we show how the principles of behaviour modification are used, in practice, to improve safety behaviour, by describing an actual application.

4

THE BEHAVIOUR-BASED APPROACH IN PRACTICE: FIRST EXPERIENCES AND DEVELOPMENT OF THE MODEL

The first environment in which the authors implemented a behavioural safety programme was in a moderate sized continuous process factory. This successful intervention encouraged the company to document and publish their perceptions and experiences for the benefit of other organizations. These insights also provided us with valuable feedback for use in the further development of what has become known as the UMIST model. In this chapter we present their views and impressions and then describe how our original model has developed with the lessons of experience.

Client Perceptions and Experience of the Implementation of the Approach

First, we examine the company's reasons and rationale for embarking on this new approach to safety improvement.

A need for change

The company, which was at the time part of an international chemical company, manufactures cellophane transparent film and has been in production for over 50 years, with a record of long service amongst the 500-plus employees. In 1992 sales were approximately £53 million. The basic process had changed little since the 1950s and was considered 'low-tech' compared to other types of film manufacture. In common with many organizations, they had seen cutbacks in demand and employment and countered these problems by making strategic changes. This included a change in focus towards speciality products and ensuring quality, service and reliability to their customers.

The process of making cellophane transparent film involves potentially hazardous, corrosive chemicals such as concentrated acids, alkalis and flammable solvents. Employees worked a continuous three-shift pattern and many personnel were required to operate large, high-speed web-fed machinery or cutting equipment, as well as a variety of other machines.

Working at heights was common and the environment was noisy and hot. There had been a long-term commitment to safety management that pre-dated the Health and Safety at Work Act. Full-time safety advisers and fire officers were employed on the site and welfare facilities included a surgery staffed by an occupational health nurse during the day. First aid cover was provided, in the first instance, by employees trained to Health and Safety Executive standards. The company claimed that approximately 20 per cent of its capital expenditure had been safety-related for many years previously. As part of its strategy of continuous improvement the plant had been accredited to ISO 9000.

Typically of such companies, many of their safety improvement efforts prior to 1991 had been reactive and in response to incidents that occurred on site. The basis for accident reduction before the association with UMIST took the form of safety committees, exhortation to do better, occasional disciplinary cases, a senior management safety policy team and senior manager hazard audits. 'Safety' was moved to the first item on meeting agendas. Whilst the number of accidents had reduced slightly year-on-year, the effort expended was not expressed in results that were deemed to be acceptable. Indeed, maintaining improvement was a problem and certain situations were acknowledged as unsatisfactory. For example:

- As expected, new safety campaigns did have some initial impact but the improvements were either not maintained or were disproportionately small in comparison to the efforts and expenditure involved. One of these initiatives involved a system of hazard spotting by safety committee teams, but the impact of this was minimal.
- A senior management safety policy and support team was set up to provide support for the safety committees. This was only partially successful because it became too involved in detail and simply created another level of safety management.
- It was acknowledged that the engineering group was not always able to meet the demands on it for safety problems; real and perceived. Shop-floor personnel interpreted this as a lack of priority concerning on-site safety needs. In fact, the company policy and message of 'safety first' was perceived, to a certain degree, as merely 'lip service'.

Whilst these dissatisfactions were very apparent among the plant's own management team, the final impetus to seek a new approach came from the parent company's insistence on improvements in the safety records of its plants. Long-term safety targets had been set for each of the group's organizations and lost-time accidents resulted in cost penalties in centrally allocated insurance premiums, and central health and safety overheads. The plant's profitability and safety record during the late 1980s compared unfavourably with other sites and by 1990/91 the plant had the third worst safety record in the group. Two very serious accidents brought the situation to a crisis. One operator had both feet crushed by a conveyor and a finishing machine operator suffered a severely fractured arm. Thus the search was on for a vehicle that would give a significant improvement in their accident record and the Operations Improvements Manager was given the task of finding a new approach to safety management.

The search for a new approach

The Operations Improvements Manager's research led him to discover an American safety management package that focused on behavioural changes. Whilst the uniquely American style of this package was not felt to be appropriate for British shop-floor personnel, the underlying principles were appealing. He, therefore, began by asking employees whether they were aware that, on certain occasions, they were working unsafely, and if so, why? These discussions revealed that employees believed that, to serve the interests of production, a certain level of accidents was inevitable. Furthermore, the workforce also believed that this was an acceptable standard, tacitly held by management, however much they seemed to publicly state the contrary view. The management message of, 'safety first, every time' was simply not believed!

However, the Improvement Manager had an opportunity to test out his belief in this new behaviour-based approach while trying to help a shop-floor worker who had suffered numerous injuries in the past, and was again injured as a result of carrying out a task in an unsafe manner. Rather than resort once more to disciplinary action the individual was required to take part in discussions about his potentially unsafe versus safe job-related behaviours. Once agreement had been reached on the desirable safe behaviours the worker was continually reminded of these, by his shift supervisor, several times during each shift. This individual remained accident-free over the following year.

Nevertheless, a more structured and time-efficient system was needed in order to introduce such an approach to the whole workforce. It was around this time that two of the current authors published an article in the RoSPA journal, entitled, 'A fatal inversion' (Makin and Sutherland, 1991). The authors described the weaknesses inherent in the traditional approaches to safety improvement and the underlying theory of a behaviour-based approach. As a result of this, and after several meetings, it was agreed that it would be appropriate to involve the UMIST team of occupational psychologists in a safety improvement initiative for the company and that the focus would be on:

- Safe behaviour rather than attitudes, and this would lead to fewer accidents; any change in attitudes is a consequence and not the cause of safe behaviour.
- Reinforcement of safe behaviour rather than the punishment and discipline of unsafe behaviours.
- Measuring safe behaviour not hazards.
- Accepting that the best people to identify safe behaviour are the operators.

- Searching past accident records and near miss records for a repeating pattern of unsafe behaviour.
- Training operators to observe and measure safe behaviour, provide visual feedback and praise to reinforce the desired safe behaviours.
- Enabling the workforce to set its own goals and observing its own behaviour.

The initiative became known by the acronym TABS – 'Think and Behave Safely'.

Putting the theory into practice

Before any practical steps could begin, certain diagnostic procedures were followed. These included the implementation of a safety climate survey, one-to-one interviews and an analysis of accident records.

- A safety climate questionnaire was circulated throughout the factory to gain views on a variety of aspects associated with safety culture and the working environment. The survey produced a 70 per cent response rate and the results helped to guide and structure the implementation of this initiative.
- A series of in-depth semi-structured interviews was conducted with a structured sample of 70 people (15 per cent of the workforce) in order to gain their views on the safe way to do their job. These interviews aimed to reveal minor accidents that were thought to often go unreported, and the behaviours that caused them.
- An analysis of the previous 24 months of accident records was carried out to identify patterns of unsafe behaviours.

After these diagnostic, information-gathering steps, the practical work could start, as follows:

- **Produce departmental checklists**: Checklists of the behaviours most commonly associated with accidents were generated from the above diagnostic techniques. These first-draft checklists were refined in consultation with departmental managers and safety committee members. Table 4.1 gives an example of one of the final checklists produced by the casting 'day gang'.
- **Train the observers**: A total of 48 observers was needed to observe across all departments and all shift patterns. They attended a two-day training programme in using the checklists and monitoring the behaviour of their peers. The first group of 'volunteer' observers was 'selected' to include safety committee members, some supervisors and some people known to show an interest in safety. Day one of the training programme included the theory and principles of the behavioural approach, which emphasized reward rather than punishment and the use of praise and feedback as reinforcement. The training also covered such techniques as

Table 4.1 *Day gang (casting) final checklist items*

Item	Safe	Unsafe	Not Seen
1 Safety glasses/goggles must be worn when: (a) cleaning coaters (b) cleaning out hopper (c) opening any valve (d) cleaning film from dryer basement (e) using power hose on acid baths (f) handling any chemicals (g) handling urea (h) making up wax			
2 Gloves must be worn when: handling dyes, bleach/maleic acid and anchoring resins			
3 Safety footwear must be worn			
4 When adjusting jubilee clips, screwdrivers must be used, not knives			
5 Open hoods must be fully raised (check to see if the hoods are raised as fully as the ones next to them)			
6 Waste film/material must not be left on the floor. Particularly around the wet-end and the winders (look for viscose on floor by jets)			
7 Steps and ladders must be descended one step at a time			
8 No jewellery should be worn (rings, watches, necklaces, etc.)			
TOTALS			

were needed to run the sessions where safety targets would be set. On the second day of the training the observers where taught how to use and score the checklists and practised with the checklists on the factory floor. The lists were modified as necessary on the basis of these practical experiences. Valuable lessons were learned at this time. For example, the observers found that the use of clipboards on the shop-floor triggered apprehension and hostility in the workforce, and so they refrained from using formal clipboards. Also, it was clear that communications to the workforce about this new initiative had not been successful and the arrival of trainee observers on the shop-floor was greeted with much suspicion and hostility. It was not a good start!

• **Setting the baseline measure**: Four weeks of observations followed the training period in order to establish a baseline measure. During this time the checklist was used on every shift. Whilst the observers were slow with the initial observations, with practice they were able to complete the checks, on average, in about 10 minutes. The data was gathered centrally and used to calculate a baseline figure for safe behaviour. This was expressed as the number of safe behaviours out of all the behaviours observed, and took into account the percentage of behaviours 'not seen'. Enlarged copies of the checklists were displayed in all departments.

• **Goal/target-setting for improved safety performance**: at the end of the four-week baseline period all employees attended a small-group, 'goal-setting' session (these were usually 'by shift' groups). At this time the observers, aided by the UMIST consultants, reminded everyone about the principles of the new approach to safety improvement and how it worked in practice. Reminders were given about the duties of the observers and how the checklists were used and scored. It was made clear that no individuals could be identified from the observations, and that no disciplinary action would be involved as a result of observer duties and actions. The group was invited to guess how safe they felt they had been during the past four weeks. These estimates varied widely and tended to be rather optimistic. The actual baseline observations were then announced and the group discussed and agreed on a target for safety improvement that was 'difficult yet achievable'. The targets set for the different shifts were averaged for each department and entered onto the departmental groups as a solid line. Table 4.2 shows the baseline average and the goal level set for each department.

Table 4.2 *Safety behaviour baseline averages and target levels set, by department*

Department	Baseline average	Goal level
Casting	27	70
Coating	69	80
Viscose and day gang	65	80
Engineers	46	85
Finishing	28	65
Reel wrap	52	75
Sheeting	100	95
Works laboratory	58	85
Power house	64	95
IT/Sales/Accounts	67	100
Engineers office	67	85
Production office	89	95
Customer services	100	95

The goal-setting sessions followed the 'SMART' principles of participative goal-setting. These are as follows:

1 Specific: that is, if the average baseline measure for the four-week period was 54 per cent safe, then the group had to specify a target that they believed they could reach, for example, '88 per cent safe'.
2 Measurable: the goal must be stated in measurable terms rather than some vague generalizations about 'trying to behave in a more safe way'.

3   Achievable and realistic: the target for improvement should be high
enough to be a challenge that is valued by the group and perceived as
a worthwhile effort. However, if it is not viewed as 'realistic', the
group will become demotivated and demoralized. Likewise, a target
that is regarded as 'too low' will not generate interest or effort.

4   Time-bounded: a date by which the target must be reached is agreed
so that the group can work towards the specific goal for safety
improvement rather than have some undefined time limit. This means
that the feedback charts and graphs can be prepared for the agreed
time span, for example 12, 16 or 20 weeks, and thereby act as a
continual visual reminder. In this instance the observers were
recruited for a total of 20 weeks, namely a four-week baseline period
and a 16-week intervention period. At the end of this time they would
hand over to newly trained observers for the next phase of the TABS
programme.

- **The intervention**: thus, the observers continued their observations and
every Friday the results were entered onto the feedback graph. Figure 4.1
shows the safe behaviour performance, by week, for the total plant. These
charts were enlarged to poster size and produced in bright colours
(yellow boards with the target line drawn in red). Failure to reach the
target is not punishable, but instead, feedback and praise are used for
what has been achieved. This also stimulates discussion on why targets
have not been met and what can be done to improve the situation. Thus,
the company found that observation by peers gained the greatest sense
of ownership and involvement, and that feedback reinforced the desired
behaviours.

- **Beyond phase one**: at the end of the first phase new observers were
trained, who continued observations with a modified checklist. Some
behaviours were removed from the checklist (on a temporary basis, if
required), if the observers had continually recorded this as a 'safe
behaviour' over a period of time. This meant that additional 'new' items
could be added to the checklist. By the end of phase two onwards, the
company were running their own programme, with some initial support
by the UMIST consultants. At the phase eight point, every department
had its own checklist of between six and 20 items and observations rarely
took more than 15 minutes. Each department had its own large score-
board on which the weekly result was displayed together with the target
line (see Figure 4.1). Enlarged copies of the checklist were also displayed
near this feedback chart. At the end of each phase the company held a
lunch for the observers and gave mementoes, such as pens and baseball
caps, that helped to maintain the profile of the programme. Additionally,
on four occasions over the first two-year period, small safety related
awards, for example safety torches, kitchen fire extinguishers and smoke
detectors, were given to all employees in respect of randomly selected
landmarks. This practice was endorsed since it is in accordance with the
theoretical principles of 'variable ratio reinforcement'.

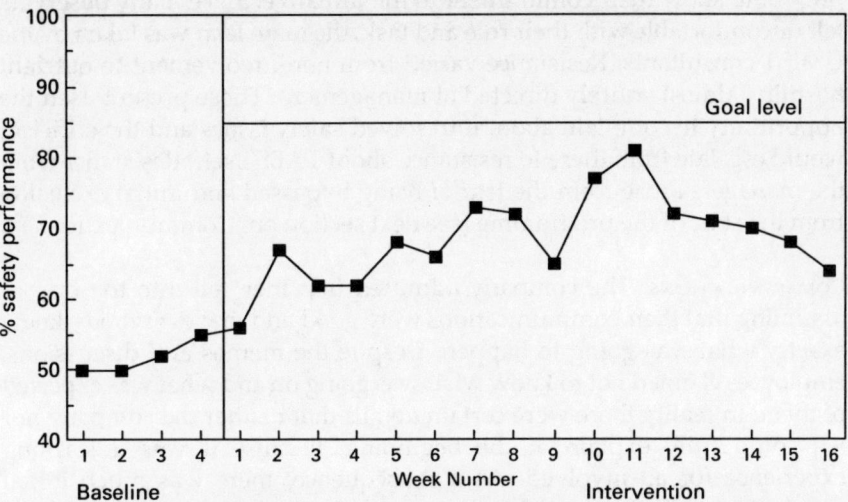

Figure 4.1 *Example of safe behaviour performance chart*

Not all plain sailing – overcoming the difficulties

The introduction of the programme was not trouble-free and certain lessons were learned. These problems were both organizational and practical, and included the following.

RESISTANCE TO CHANGE The introduction of this approach to continuous safety improvement was dependent on a high degree of shop-floor participation and had an impact on the process of management. Therefore, resistance to change was encountered at all levels. The use of UMIST staff as consultants had both negative and positive connotations within the company. Employees were very suspicious about consultants because there had been an occasion in the 1980s when the arrival of consultants on the premises resulted in large-scale redundancies. However, the UMIST consultants were perceived as academics who were regarded as independent and objective. Whilst the workforce were somewhat cautious about the presence of occupational psychologists on site (referring to them as 'the shrinks'), early resistance among the observers was overcome during the training period and while the baseline measures were being taken. Only five of the 48 trainees felt unable to act as observers during phase one of the programme, but subsequently took their turn at later stages in the programme. It was vital that the principle of anonymity and confidentiality was not breached during these early stages. Also, management had to adhere to the guarantee that no employee would be identified from a completed safety checklist and punished for behaving in an unsafe way.

The most forceful resistance came from the other operatives and tradespersons during the target-setting meetings. The original intention was that the observers would conduct these meetings and that managers would be present to show their commitment to the initiative. Since many observers felt uncomfortable with their role and task, the main lead was taken by the UMIST consultants. Resistance varied from non-involvement to outright hostility, almost entirely directed at management. Those present used the opportunity to complain about unresolved safety issues and the criticism would escalate from there to resistance about TABS itself. Resistance from the managers arose from the fear of being bypassed and initial exclusion from the start of the programme (see next section on 'Communications').

COMMUNICATIONS The company admitted that they fell into the trap of assuming that their communications were good and that everybody knew exactly what was going to happen. Despite the memos and discussions, employees claimed not to know what was going on and what was expected of them. In reality there were certain details that neither the company nor we could fully explain at the beginning, because it was a learning experience for all involved. As a consequence, there was much initial suspicion and distrust and a feeling among both observers and observed that this was a spying exercise for management. Indeed, they believed that discipline would eventually catch up with those who were seen not behaving safely. Some observers withdrew at an early stage and others had a rough ride initially. The central tenet of the scheme – the devolvement of responsibility for safety to those most affected – requires willing and active participants. The early problems in communication led to this principle being compromised. UMIST had asked for volunteers to act as the first observers, but discovered during the training course that the trainees had been instructed to, 'attend a course on safety'. Few of these 'volunteers' had full details of what the course entailed or what was expected of them! Therefore, their apprehension was justifiable. They had been 'volunteered' to take part in a programme lasting 20 weeks, which required them to 'take more responsibility, do more work (including paper work) – all without any extra pay'. Inevitably, the consultants had a difficult task in explaining what was in it for them and why they should give this new approach a chance of success.

Another factor was that not all managers and supervisors felt that they knew enough about the programme. Some had attended a briefing session run by the consultants, but others were (or should have been) briefed by their own bosses. Clearly this proved not to be enough to help them support the programme and the observers. This oversight was quickly rectified by putting operational managers and supervisors through a half-day training programme. As part of their training, managers were asked to assess how much they used 'praise' in relation to safe behaviour. Typically, managers and supervisors overestimate the amount of praise given, and such was the situation at this site. The organizational culture and climate emphasized the issuing of instructions and ensuring compliance with discipline. With only partial commitment from some managers, the use of praise to promote the programme was not as great as it should have been. Therefore, as consultants, we felt that we should have retained more control over these early stages and been involved in the communication processes at the start-up of the programme.

SHIFTS The process of averaging the targets for shifts caused some dissatisfaction because it meant that certain shifts felt that their target had been influenced by the behaviour of other shifts which was beyond their control. As we often find, there was some shift rivalry and competition. It was clear that some employees believed they could improve their safety performance if they did not have to combine their scores with those of colleagues who worked on a different shift. Ultimately, in some departments they did change to target-setting by shift and each shift had their own feedback chart. Obviously, the element of competition must be monitored carefully to ensure that it does not become negative and unhealthy in its consequences.

Amending the programme

As the programme continued other issues emerged, and these were dealt with by making certain amendments to the programme. This is the main reason for our current recommendation to start the behavioural approach by introducing a pilot study into one area or department within the organization. In this way, many of the situations described below could be dealt with on a small-scale, at a low threat level. We recommend two criteria for choosing a pilot area. The first is that it should be possible to show an immediate and visible improvement to safety so that other people in the organization can see that it is a worthwhile initiative. Secondly, we suggest the selection of a group of employees who are willing to be the 'guinea pigs' and keen to test out something new. It is important to remember that the steps to implementation that we outline are merely guidelines. The programme must be tailored to fit the organizational structure and climate and the needs of the workforce, if it is to succeed and endure. Nevertheless, as we have already stated, this initial project was a learning experience and the following situations that led to amendment of the original programme came from the company's feedback.

- **Lack of initial involvement**: Although the safety behaviour checklists were circulated to managers for comment, it was only some time later that it was pointed out that the inclusion of mandatory items, such as the removal of all jewellery, caused problems. Non-compliance with these well-established items was a disciplinary offence. Inclusion of behavioural items of this type on a checklist conflicted with the 'no discipline' promise given with the checklist. Such items were subsequently removed.

- **Use of 'volunteer' observers**: The company had initially decided that only volunteers would be trained as observers. After the first two phases, the feedback from the observers was that everyone should be trained. This was to ensure that the message was spread across the whole workforce. Although no one was forced to be an observer, by the end of phase eight, 355 employees (90 per cent of those eligible) had been trained as observers. At the two-year mark, only a few people were still not part of the programme. For some employees this was a concern because of literacy problems, whereas others felt 'too old' and close to retirement to become involved in learning something new. The company resolved this by insisting that these individuals (that is, less than 10 per cent of the workforce) sat in on a one-day familiarization TABS course in return for being excused observer duties. Eventually, when everybody had been through the programme, the process began again. Observations and feedback had become a way of life.
- **Target-setting meetings**: Since the five-shift system restricted the opportunity for departmental meetings and it proved to be too difficult and costly to get enough employees together for the goal-setting meetings, the original, formal target-setting meetings were replaced with an informal process whereby the observers consulted with their peers at their workplace.
- **Data collection**: Originally the data were collected on pro-forma sheets and returned to a central point. Eventually a computer program was written which enabled the observers to enter their results directly on to the shop-floor terminals of the site's computer. Thus the calculations and results could be generated quickly and accurately. Access to the information was improved and a data bank of past phases led to an improvement in checklist generation and feedback. For example, it was possible to identify specific items on a checklist in terms of 'most safe' and 'least safe' behaviour. In this way, by discussions and problem solving sessions, the workforce could concentrate their efforts on the behaviours in most need of safety improvement. These were produced as bar charts and displayed, in colour, beside the feedback charts (see Figure 4.2).
- **Checklist modification**: The critical safety behaviour checklists are modified as new areas of potentially unsafe behaviour are recognized. All employees are encouraged to identify items for inclusion on the checklists.
- **Using observers as 'trainers'**: Eventually some observers began to take over part of the training role. Certainly the practical aspects of being an observer can be carried out as a 'Sitting with Sid or Nellie' exercise. The new trainee observer has the benefit of the experienced observer to act as a mentor and support, and to share experiences. Of course, these observers can also be available to take over the observation duties if a co-worker is absent from the workplace, or working an extremely busy shift, which prevents them from going out to do their observations.

Safety Figures – Coating: w/e 16 July

Item

1	Gloves when handling film cores
2	Eye protection between doctor rollers and chill
3	Worn razor blades in bin
4	Hoists free running on trackways
5	Fibre glass cores dragged on floor
6	Waste film around towers
7	Earthing clamps filling solvent buckets
8	Solvent bucket lids on
9	Unwind hoist straps flat and 2" from core
10	Wind-up hoist securely located on core
11	Masks when handling resin
12	Millroll hoist bridges securely located
13	Walking under suspended rolls in stores

Number of behaviours observed

■ Safe   ▨ Unsafe

Figure 4.2 *Example of bar chart to show safe/unsafe behaviours observed for checklist items*

Conditions for success

The following points were identified as key steps to success:

- The concept and method must be known to everyone.
- Managers and supervisors must feel a sense of ownership from the start.
- Everyone must be told about the first training sessions so that the appearance of trainee observers on the shop-floor does not lead to misunderstanding, suspicion or hostility.
- The training must be appropriate to the organization, plant or factory.
- Observers must not become invisible and steps must be taken to ensure that both they and the initiative maintain high visibility and high credibility in the company.
- The unions must be involved. The unions on site were involved in the programme and some safety representatives acted as observers. Other safety representatives feared a conflict of roles and were apprehensive that their ability to defend a safety disciplinary case, or represent an employee in an accident, could be compromised. Nevertheless, the general feeling was that any increased workforce involvement must improve industrial relations. However, we do add a caution at this point, in that an organization should be aware that increased levels of participation by the workforce in one area will inevitably raise expectations elsewhere. If this is managed well it can only be beneficial to the organization and the self-development opportunities for employees.

- It is likely that an external consultant will be needed in the early stages of the programme. However, experience has shown that a company can quite quickly become independent if it chooses.
- The project must have a 'champion', usually a management figure. It is important that the observers and managers do not become totally reliant on this person. Also, it does not mean that only one person in the organization is responsible for the safety programme and that other managers can get on with something else! Although it is a 'shop-floor' driven model it is important that managers and supervisors treat this as any other operational task within their remit. They need to support, guide, encourage and show a genuine interest in the activities of the observers.
- The project must have the commitment of senior management. The company had a new company chief executive who worked hard to improve communications. His participative approach and accessibility influenced the process of change that was already under way. He was highly visible on the shop-floor and actively showed his commitment and support for the observers and the TABS programme.

Measuring success

The majority of employees who took part in this new safety improvement initiative were convinced that it is was a success. For example:

- In a typical 16-week period in 1991, prior to the start of TABS, the company had 118 accidents, including all minor first aid treatments. At the end of phase one in 1992, there had been only 63 similar incidents. By the end of phase twelve, four years later, this figure was only 26 incidents in the 16-week period (see Figure 4.3).
- In the first two years, three-day lost time accidents (LTAs) fell by 86 per cent and minor accidents by 72 per cent. The number of LTAs fell from 22 in the year prior to the introduction of the programme (to March 1992), to six in the following year and to three in 1993/94. The LTA rate was 11 per 1,000 employees for 1992/3 and 5.8 in 1993/4. The parent company had set a target of 5.3 for the site by 1995/6. The company's position for LTAs within the group was 18th out of 21 in 1991; by 1993 it had risen to 10th. This was at a time when all group sites were striving to improve. When the DuPont Consultancy Service carried out a world-wide survey of the parent company sites in 1993, the plant came joint second for safety procedures and the commitment to put them into practice.

Not all the improvements observed were quantifiable, but the company did acknowledge other benefits stemming from the improvements in communication, involvement and commitment that the TABS initiative produced. A sense of achievement and pride became manifest as a greater

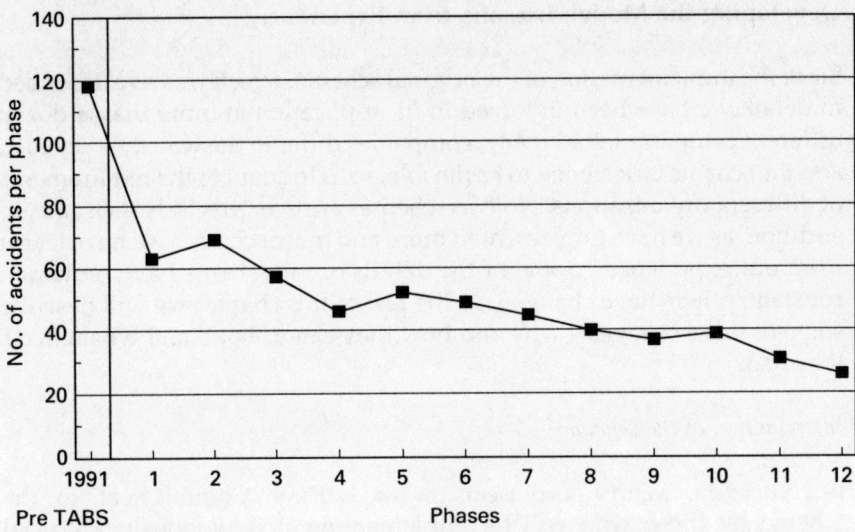

Figure 4.3 *Accident rates in the intervention plant, 1991–1996*

confidence and initiative among the workforce. When the company started preparing its risk assessments it was able to find employees on the shop-floor who were able and willing to do the work. Specialist risk assessment safety committees are able to carry out assessments for each operation, based on their on-the-job knowledge. The defunct 'hazard-spotting' groups came forward and asked to start this activity again. They organized this in their own time and generated their own timetable and programme. A 'suggestion box' was started up again and was actively used. The 'near-miss' accident reporting scheme was revived with considerable success. Prior to the start of TABS, management complained that nothing was ever reported, and employees responded by saying that nothing was ever done. It was an impasse! Increased cooperation led to improved credibility for the scheme. Simple changes led to significant improvements: for example, a report form was produced that had a detachable slip that was returned to the originator to show what action had been taken (what and when, etc.). This was eventually added to the computer system. Now any employee can file 'near miss' reports, suggestions and hazard-spotting situations directly at the workstation terminal and gain full access to feedback on the progress of his/her report.

The company has also successfully used the behaviour-based approach to improve meetings. However, the ultimate success came when employees realized that the model used in their safety improvement programme could be directed to their quality improvement initiative. (This venture is described in Chapter 7.) As the Operations Improvement Manager said, 'Is it the answer to all our ills? Of course not, but it worked for us and it could be worth finding out if it could work for you.'

Developing the Model: Lessons from Experience

Since the implementation of the original scheme, which we have described in detail, we have been involved in its application in more than a dozen different companies. Inevitably, companies differ in the way they operate, and the scheme is designed to be flexible, so as to adapt to the requirements of different organizations. No two schemes are ever precisely the same. In addition, as we have implemented more and more schemes we have learnt from our experiences. Some of the details of the scheme have remained constant, others have changed. In the rest of this chapter we will describe some of these changes – why and how they came about, and what effects they had.

Introduction of the scheme

INITIAL PRESENTATION AND DECISION ON SCHEME Commitment to the scheme by those who will be implementing it is obviously a crucial element in the likely success of the scheme. Our experience is that the first batch of observers is likely to be drawn from those with an existing interest in safety issues, trades union and shop-floor safety representatives, for example. Often these individuals will be members of safety committees, and it is often these committees that are the most appropriate starting point for a scheme.

We are now very used to doing presentations to such committees. These presentations are intended to highlight the twin central components of the scheme. The first is that the ownership of the scheme rests with those who are to be affected by it. The second is that the use of discipline is ineffective in trying to change people's behaviour. To this end, a brief explanation of the underlying behavioural theory is presented. In our experience, the most common climate of such presentations is one of healthy scepticism. This is perhaps to be expected. We have never experienced such a situation, but we have often said that we would be unlikely to recommend that an organization goes ahead with a scheme in a climate of outright hostility. The decision as to whether to proceed with a scheme should rest with the safety committee. In our experience, it is very dangerous to try to use a committee as a vehicle for endorsing a decision that has, covertly, already been made elsewhere.

GAINING ACCEPTANCE OF THE LINKS BETWEEN BEHAVIOUR AND ACCIDENTS During our initial presentations in some organizations it became clear that certain members of the workforce found it difficult to accept and understand the statistical fact that 90–95 per cent of all accidents are, in some way, attributable to human behaviour or human error. Much of the hostility and resentment was associated with feelings of 'being blamed' once again. It was apparent that the workforce believed that we, as consultants, were giving the presentation at the request of management, and that this must therefore be a ploy on the part of management to place the blame for a poor accident record on employees, thereby avoiding taking further action themselves (or, more particularly, spending money on safety improvements!). While we were able to allay many of these fears during the presentation and the training programme that followed, one company produced an innovative and interesting approach to raising awareness on the links between behaviour and accidents. This included a session specifically on accident causation with the use of a video exercise.

These sessions begin with an open floor debate which aimed to identify the conditions or situations that would increase vulnerability to accident involvement – not just at work, but also when in the home or when driving a vehicle. For example, physical conditions, such as being too hot, too cold, or too stuffy and humid, can impair concentration levels, slow reaction time and cause a feeling of fatigue. Exposure to a work environment that is too noisy, poorly ventilated, unclean or smelly can also affect performance in a negative way, since it takes up attentional capacity, or impacts on hearing or vision. In addition to these conditions, there are a wide variety of factors in the work environment that are potential sources of stress. For example, the need to work shifts, including night shifts, exposure to a very high workload, lack of control over machine pace and working with difficult colleagues or a dictatorial boss are just some of the sources of stress at work which can increase vulnerability to accident involvement. Many people fail to realize that they can make this situation worse by resorting to strategies that are liable to make them less able to cope in the long term *and* more likely to become an accident statistic. These behaviours include alcohol and drug abuse (and nicotine from cigarette smoking), the excessive intake of caffeine and the addictive behaviours linked to the abuse of sleeping pills, tranquillizers and anti-depressant drugs. Ultimately, these can render the individual anxious, depressed, obsessional and fatigued, which can affect levels of self-esteem or job and life satisfaction. Therefore, it is no wonder that accidents are more likely in such conditions!

The second part of the session involved a group exercise. After watching a brief video of a garden scene in which two friends were working, the group had to discuss the scenario and identify any actions which were unsafe for the individual and those that were unsafe for other people working in the garden. They also made a list of the behaviours that were required to make the garden work safer and how the gardeners could be encouraged to change their behaviour to work more safely. In a plenary discussion the groups were asked how these 'garden-related' behaviours might apply to unsafe acts in the workplace, and what stops us from encouraging each other to be safe. The unsafe actions identified were also plotted onto a 'Heinrich's Triangle' (see Chapter 1). In this way it became clear that there had not been any deaths, no lost time accidents, only two minor first-aid events, but at least 15 unsafe actions performed by the two men that could have resulted in an accident. At least two of these were potentially fatal. Thus, in this way the group could clearly see the links between behaviour and accident occurrence and that accidents are not usually due to any deliberate, intentional or malicious acts of negligence. Most of the incidents occur because people develop bad habits and do not realize they have become vulnerable.

Without doubt, the use of this innovative addition to the basic training programme contributed greatly to the success of the programme and the high level of commitment and ownership demonstrated by this workforce.

SCOPE OF SCHEME In the original intervention, the decision had been made to introduce the scheme throughout the whole plant, including the offices. Experience has taught us that it is often preferable to introduce the scheme as a pilot into one area, and then gradually extend it to others. No two schemes are precisely the same and the use of a pilot area gives the opportunity to 'fine tune' the system to the precise needs of the particular organization.

There are a number of factors that need to be considered when choosing a pilot area.

- Ideally, detailed accident statistics should be available, as these can form the basis for the generation of the initial checklists.
- It should be an area where some improvement might be expected. Although it need not necessarily be the area with the highest accident figures, it should not be one with a very low level of accidents. As with any new scheme, early successes are always desirable.
- The area should have a relatively stable workforce. Trying to introduce any new scheme at a time when a large number of new workers are likely to be inducted into the organization is to be, if at all possible, avoided.
- Ideally an area should be chosen that is clearly identifiable as a coherent 'unit'. This is usually not difficult to do. Most workers can, when asked, identify by name areas with which they identify and which are run by 'natural' working groups. These may be departments or sub-departments.
- What also needs to be considered is the number of observers who will need to be trained. The area should be capable of being divided up into natural observation 'areas'. The optimal size should be such that an observer can walk around the area, taking observations, in about 10 or 15 minutes. In order to provide cover for things like sickness and holidays we have found that it is desirable to train two observers for each area. (In the case of shift-work there will need to be two observers trained for each shift.)

COMMUNICATION ABOUT THE SCHEME Whether or not the decision is made to undertake a pilot project, the whole of the workforce should be briefed as to the nature of the scheme. This is usually a three-stage process.

1 Management letter: as we have found, management commitment is an important element in the success of any scheme. The initial introduction of the scheme must, therefore, be seen to have the support of the senior

site management. This commitment is best communicated, in our opinion, by a number of methods – the first of which is often a letter from the senior management to all employees. An example of such a letter is given in Figure 4.4. The level of detail in this letter will depend on whether the workforce briefing sessions are already in progress. Without a doubt, communication is best when it is in more than one medium and the message is repeated. The axiom, 'say what you are going to say; say it; and then say what you have said' cannot be emphasized enough.

2 Newsletter: the original letter may also be reinforced by inclusion of details in any company newsletter. This may include descriptions of where the scheme has been successfully introduced elsewhere, either in other plants within the same group, or other external plants.

3 Workforce briefing: this briefing should be similar to that mentioned above. The content of such briefings is, in our experience, soon forgotten by those not directly involved. Nevertheless they have the function of keeping the workforce informed and thus helping to prevent the spread of rumours that are often wild and inaccurate. It is often useful if these briefings are introduced by a senior manager. This again indicates management commitment to the scheme.

CHOICE OF OBSERVERS As mentioned above, the first observers are often drawn from those on the safety committee. Ideally they should all be volunteers but, in our experience, they are often 'nominated' volunteers. Some, therefore, have 'volunteered' more than others. Whilst some are enthusiastic from the start, others sometimes feel 'token' resistance is required, if only to demonstrate their independence. It may be that the 'voluntary' nature of the scheme is being tested, to see whether refusal is followed by what may be perceived as more coercive methods of persuasion. Such coercion would indicate that the 'voluntary' element was merely a veneer. Coercion should, therefore, be avoided. In our experience enough volunteers eventually do come forward.

Whilst we prefer the use of the phase, 'volunteer observer', in reality this should be expressed as 'taking one's turn to be a volunteer observer', since one of the important features of the scheme is that everyone does a stint as an observer. Indeed, it is often difficult for an individual to put themselves forward as a willing volunteer when they are keen to take part in the programme but their co-workers are not so willing. The only reward we offer is that of working in a safer environment and, on first impression, this, for some employees is not enough in exchange for the extra duties and responsibility. Many view this as extra work without extra pay. So, the true volunteer observer can experience conflict. However, this can easily be overcome if he or she is able to tell their workmates that they have been 'asked' by the supervisor or line manager, to act as an observer. We would not support the payment of observers for these extra duties since 'safety in the workplace' should not be bought. In addition, employees, while actually, taking part in observer duties are excused from their normal operational tasks.

To:

From: CEO or MD

Subject: Improving safety performance – a new approach for Company XYZ

Date:

With the help of [insert name of consultants], who have recently worked on similar projects [where], we have initiated a project to further improve safety at work. This approach is different because it is employee owned and employee driven. [You might want to tell them more about the consultants at this point.]

The objectives
To continuously improve safety performance by encouraging safe working practices across the site.

The approach
The approach is based on the following ideas:

* People doing the job know most about day to day safety.
* It is more effective to concentrate on behaviour rather than personality or attitudes. Most safety improvement initiatives focus on changing people's attitudes because of the assumption that attitudes cause behaviours and can be changed. However, there is considerable evidence to show that attitudes have an uncertain link with behaviour. So – this approach just focuses on behaviour.
* Emphasis on the encouragement of safe behaviour is more effective than the use of sanctions or discipline.
* After the measurement of levels of safety, targets for improvement will be agreed by the workforce.

But, this approach does not replace any current safety initiatives in existence, e.g. hazard and near miss reporting procedures.

A need to improve
[Include something in here about your current performance/unsatisfactory state etc. and the need to involve everyone in the improvement of safety performance.] This approach has been very successful in other sites similar to ourselves and has a good 'track record'.

[Here give details of what will happen and when: the level of detail will depend on whether the briefing sessions have already begun.]
We are starting with a pilot project in order to work out the best way of implementing the approach in Company XYZ. Employees from [which departments, etc.] will act as observers initially. After training they will take observations on each shift, using a checklist, constructed by personnel for their own area, and score each checklist item as 'safe' or 'unsafe'. At the start they will observe to obtain a baseline measure of safe working performance which is recorded as a percentage – that is, how safe are you at the moment? This takes four weeks. When this is established you will be able to agree a target/goal for improvement. Throughout the observations, which will run for 16 weeks, the observers will give you feedback on safe/unsafe behaviour, and together you will agree about the accepted ways of working safely. Remember, the observers are only recording information which is used to make the workplace safer.

Each week you will see the results on a feedback chart. The aim of the project is to improve the percentage of safe behaviours observed. All the information recorded, although anonymous, is public and observable.

How can you help?
Soon, you will be asked to contribute some ideas for checklists and to comment on the draft lists as they are prepared. Also, we ask that you support and help the observer to fulfil their task which solely aims to improve your safety at work on this site.

Anonymity
Your colleagues will make regular observations in each department/area and complete checklists. Copies of checklists used will be displayed in the department. The nature of these checklists and the way in which they are scored ensures that it is impossible for individuals to be identified.

Timing of the project
[When will programme start, training dates, etc.]

The Success of this Project Depends on Your Support and Cooperation
Safety in the workplace is in everyone's interest and so everyone on site in the pilot areas will take part in this project. This is your opportunity to participate in a new venture which aims to make this a safer place in which to work. We ask that you support your colleagues, the observers, and work together to encourage safe working behaviour.

Thank you for you cooperation

SIGNED ....

Figure 4.4 *An introductory letter from management – some ideas*

One problem that occasionally occurs concerns a small number of people who are totally resistant to becoming an observer. The line we usually take is to say that the decision as to whether to force them to become observers rests with management. In an ideal world, all the observers should be volunteers. The problems that can arise from having a less than committed observer are fairly obvious. Our view, therefore, is that it is legitimate for management to insist that everyone attends an observer training course. Often, attending such a course may allay fears and lead to the person agreeing to become an observer. If, however, at the end of the course they are still totally opposed, we would not force them to become an observer if it were against their will. The underlying reason for such opposition may not always be immediately apparent. We have had workers whose resistance has been due to them having very poor literacy skills, which would have been highlighted during training. (How common this is we do not know.)

The most commonly reported experience of observers is that they are treated, at least initially, as 'management spies'. This is understandable. Although the workforce will have been briefed that the central element of the scheme is the encouragement of safe behaviour, and that discipline has no part to play, old attitudes and expectations die hard. Such expectations and attitudes are often laid to rest by the observer training courses, but it is obviously impractical to train the whole workforce in one go. Until all the training has taken place, management must do what they can to support the observers, especially by ensuring that over-zealous individual managers do not resort to thinly disguised criticism and discipline when the results are not as good as they might have wished. Acknowledgement and praise for the efforts of the observers and the continual reporting on the progress of the programme help the workforce to feel less threatened and more acceptant of their new responsibilities.

Training

Since the original scheme, we have contemplated two major changes to the training courses. One we carried out; the other, after reflection, we abandoned. The change we considered, but decided against, concerned observer training; the one that we carried through concerned the training of managers.

OBSERVER TRAINING   The behavioural approach to safety can be seen as a technique based on a theory. At one point we had almost come to the conclusion that the technique could be taught, and implemented, without a knowledge of the underlying theory. We were preparing, therefore, to drop the theory part of training and concentrate solely on the technique. Two things convinced us that this was not a desirable course of action.

One was the response of the observers to the theory. Not only did they appear to find it interesting, but also it helped convince them of the non-punitive nature of the scheme. The second came as the result of a visit to an organization that had implemented the scheme without consulting ourselves.

The organization in question had visited the site of the original scheme and decided that they would copy it within their own environment. After some time they contacted us and requested that we visit them to help them understand why their scheme was not working. After only a short while it became apparent that they had not understood the theory behind the scheme and hence were making mistakes that those familiar with the theory would not have made. As will be recalled, one of the criticisms made earlier about some schemes to reduce accidents was that they encouraged people to cover up the occurrence of incidents. It transpired that, although the organization had implemented the behavioural scheme they had also kept running an existing scheme, which gave financial rewards to workers for achieving accident-free periods. It was apparent that an understanding of the theory would have helped the organization concerned not to make this fundamental mistake. We decided, therefore, that the theory should be included so that, when organizations made modifications to the original scheme, they did so in a way that did not go against the theoretical basis of the scheme.

MANAGERIAL TRAINING   The second change that we considered making concerned management training, and this was carried through.

Early in our experience we became aware that an important ingredient in determining the success, or otherwise, of a behavioural scheme was the extent of management support. We will return to this later, when we consider how schemes are implemented, but the outcome of our experiences was that we now insist that managers undertake a training course similar to that of the observers. This course covers the theoretical basis of the behavioural approach and deals with the behaviour expected of managers. It does not cover in great detail the technicalities required by the observers, so the course is normally about four hours, rather than a day and a half.

This session includes a discussion on the best hopes and the worst fears of the managers and supervisors. It is useful at this point to consider the hopes and fears expressed by the observers during their training sessions (permission is obtained from the observers to present this information at the manager training sessions). Indeed, at least one company have run their observer and manager training sessions jointly. Therefore, the opportunity to openly discuss each other's hope and fears was a very useful exercise. Likewise, the management group also produced a list of action points and this meant that these could be discussed and dealt with by both observers and management present at the same time. High degrees of trust and openness are vital for the success of joint training sessions and, in our experience, observers are very wary about the presence of anyone from the management team while they discuss their hopes and fears, generate their checklist items or identify the potential barriers to their success as an observer.

Checklists

CHECKLIST PRODUCTION   As described previously, the initial checklists are generated using data from a number of sources, including accident statistics and interviews. It is important to remember, however, that the scheme is 'owned' by those who will be carrying out the observations. The initial checklist should only be used as a starting point, which is open to amendment by those involved. What is useful, however, is to present an analysis to the observers of the accident statistics, analysed by the part of the body injured. We have found that using an enlarged diagram of a stylized human body with the percentages of accidents to each part indicated can often focus attention on trivial behaviours (for example, not wearing cut-proof gloves, goggles etc.) that may lead to a large number of relatively minor injuries. We have found that observers react well to being asked to estimate the area of the body to which most accidents happen. Often their assessments are fairly accurate, but sometimes the information comes as a surprise and so helps to provide motivation for a change in behaviour.

We have, on occasion, used focus groups to generate the first checklists but we now find, that with experience, those produced by ourselves are usually good enough as a starting point. It is also helpful to simply ask individuals to identify the items that they believe should be on a 'critical safety behaviour' checklist. Usually we distribute a written guideline for this at the end of the briefing session and, after explaining how it is be to used, we ask that these are taken out on to the shop-floor and completed during the next one or two shift or work sessions. This approach seems to be more productive than asking for the forms to be completed at the end of the briefing session. Ideas for checklist items are best generated as the individual goes about his or her daily task and has an opportunity to consult with work colleagues. Figure 4.5 illustrates one format that could be used to help the workforce think of items for inclusion on the checklist. It provides an example and categories, such as, housekeeping, the wearing of personal protective equipment, body placement, procedures, and mechanical handling and lifting as a guide. These, of course, will be determined by the nature of the job. On completion, these can be collated, by shift, department or work area and a draft set of checklist items can be produced, ready for further discussion by the observers.

Only on one occasion did a group of observers tell us that they felt they had not had enough input to the production of the checklist. However, it is a very difficult task for a group of observers to start from scratch. Thus, the analysis of accidents and the use of a pro-forma guide to identify critical behaviours seems to be the most useful starting point. We have also suggested to some clients that the safety manager, using the examples we have provided, could produce the checklist. As yet, however, no one has taken up this suggestion.

A word of caution that should be mentioned regarding checklist items. This concerns the inclusion on the lists of 'mandatory' items. The principle behind the scheme is to move from punishment to encouragement, but if items are included on the list that are compulsory (i.e. a disciplinary offence), the use of discipline would be in conflict with the stated aims of the scheme. We would strongly recommend, therefore, that such items should not be included on the checklists. We would, however, allow onto the checklists items that are 'technically' disciplinary, but where custom and practice has been that they have not been the object of disciplinary action. (If it is later decided that they should once again be enforced by disciplinary action, they should be removed from the checklist.)

Once the checklists have been finalized it will be necessary to provide a supply of them to allow the observers to record their observations. In our experience the best way of doing this is to make a supply available which can be stored in an area to which the observers have easy access. We have found that the provision of an office, if possible, provides a focus both for the scheme and the observers. We have found that, at the least, most companies will provide a lockable filing cabinet and a desk, either on the shop-floor or in an under-utilized office area.

CHECKLIST SCORING The method of checklist scoring, and the method of reporting the scores, has been described previously. It is worth, perhaps, mentioning the possibility of using a computer program to calculate and

---

Name of Company
Identifying Critical Safety Behaviours

We need your help in generating the checklist of safe behaviours. Think about your job or the area where you work, and where possible, for each of the categories below, describe a behaviour for which there is a clear distinction between 'safe' and 'unsafe' outcomes.

For example, if 'wet waste on the floor' might result in an injury caused by slipping or falling, then the SAFE behaviour would be 'floors to be free of wet waste'. Or, if the handling of chemical X is likely to lead to injury if personal protective equipment is not worn, then the SAFE behaviour would be, 'eye protection and gloves to be worn when handling chemical X. **IT MIGHT HELP IF YOU:**

A) THINK OF INCIDENTS (near misses, or dangerous occurrences) or ACCIDENTS WHICH HAVE ALREADY HAPPENED TO YOU OR YOUR COLLEAGUES

B) THINK ABOUT THE POTENTIAL ACCIDENTS THAT MIGHT OCCUR

Use the back of the form if you need extra space or wish to add items which do not fit into the categories below.

1   HOUSEKEEPING

2   THE WEARING OF PERSONAL PROTECTIVE EQUIPMENT

3   BODY POSITION/PLACEMENT

4   SYSTEMS/PROCEDURES

5   HANDLING/LIFTING (manual or mechanical)

Finally, we do not need your name, but because some items might relate only to specific activities or types of work, please help by indicating which department or area you are in.

Return your completed form to us next time you are on shift. We will use this information to help produce the first data checklists.

Thank you – Valerie Sutherland; Charles Cox; Peter Makin –   **[Date]**

Figure 4.5 *A suggested format for the identification of critical safety behaviours by the workforce*

report the results. In the original study reported earlier, computer scoring of the checklists was introduced after the scheme had been in operation about a year. We are also aware that software packages are available on a commercial basis from some consultants. Computer scoring can be useful but we would not advocate it as an initial method for two main reasons.

First, we are of the opinion that to use computers in this way is somewhat analogous to school children using calculators – they can be very useful, but they should only be used once the basics underlying them have been properly understood (that is, they are no substitute for a 'times table'). Secondly, as we have mentioned before, no two schemes will be exactly the same. Changes will be made so that the scheme fits the specific requirements of each organization. The use of computer scoring should, we feel, develop naturally from the needs of the observers. The use, from the outset, of a standard package would not allow such a natural development. The system is designed to be bespoke rather than 'off the shelf' and, in this case at least, bespoke is both better *and* cheaper. For these reasons we feel that manual scoring should be used initially, with computer scoring as a later addition, if it is felt that it would be useful.

At the start of a new initiative the workforce often fear that some punishment will eventually follow the observation and recording of an unsafe act. When the checklist procedure and scoring procedures are explained, these fears usually disappear. However, some observers and their colleagues remain anxious and insist that management do not have access to the completed checklists. This complicates the role of the supervisor and manager as a supporter of the observer and the programme. Until the message about confidentiality, anonymity and the promise of no disciplinary action for unsafe behaviours associated with checklist items is firmly understood and accepted by both employees and managers, it is best for the completed checklists to remain the property of the observer, or a lead observer. They do need to be kept for a while to assist in the updating of the new checklists. However, many organizations have found that the observers are quite happy to file the completed checklists in a central point so they can be analysed at a later date. A climate and culture of openness and trust is vital if the organization intends to transfer the checklist scoring process to an open access computer-based system.

Implementation – management support

Perhaps the most important element in the ongoing success of any scheme is, in our experience, the support given to the scheme by management. One must gauge the correct level of support carefully, so as not to fall into one of two traps. These two traps may be seen as lying at opposite ends of a continuum, with the trap of standing back too far at one end, and that of becoming too directly involved at the other. The scheme belongs to the observers and their fellow workers and too close an involvement by management may lead to that ownership being threatened. Not enough management involvement, on the other hand, may lead the observers to conclude that management do not care whether the scheme succeeds or fails. Both these extremes will lead to a loss of morale and commitment by the observers. Not only do the management need to support the scheme, they need to be seen to be supporting it!

There are a number of things that management can do to ensure that their support is not only given, but perceived to be given.

MONITORING Recently some behavioural scientists, most notably Komaki (1998), have been turning their attention to the study of effective leadership. The question they pose themselves is, what are the specific, observable, behaviours that effective leaders exhibit? The research has been very detailed, but one conclusion stands out. Effective leaders, unlike their ineffective counterparts, **monitor the behaviour of their subordinates on a regular basis**. This is not to say that they are directive or authoritarian. Monitoring need not (perhaps should not) be a matter of standing over lazy subordinates in order to ensure they work. Monitoring is rather a matter of taking an interest in the work-related behaviours of subordinates. This interest has two main effects. First, it allows the manager to check whether things are going as they should. If not, steps can be taken to ensure that corrections are made earlier, rather than later. Secondly, when things are going well it provides the opportunity for the manager to give positive feedback and reinforcement to the worker/s, an essential element for a highly motivated workforce. In terms of the scheme, such monitoring may consist of, for example, 'walking around' the feedback charts at least once in any working day – having looked at the charts, going to talk to the observers and the workforce to give encouragement.

REMOVING BARRIERS During the observer training matters often arise which the observers consider to be barriers to the effective implementation of the scheme. Many of these are what we refer to as 'engineering' problems: examples might be that hard hats are not available, or that a portable welding screen is needed. Such issues are recorded during the training on a list of 'action points'. At the end of the course we normally ask that a 'relevant' manager attends to comment on these action points. It is important, especially during the early stages of the scheme, that management respond positively to these suggestions. Some will not be possible to correct, often due to a lack of resources. But where this is genuinely the case management should not promise what they know they cannot deliver. It is generally good policy to be open and realistic when this occurs. In our experience observers are not unrealistic in their assessment of what is, and is not, possible. Many of the items, however, will be capable of being resolved. In these cases the manager should make a note of the problem, identify a person who will take responsibility for the matter, and give a precise indication of when that action will be taken.

Implementation – improving feedback

Without doubt, one of the most difficult duties for the observer is giving feedback about safety performance to a co-worker. Sadly, many of us are guilty of not giving praise or thanks for work or a service done well. We tend to be much more likely to criticize or complain when we are dissatisfied. This behaviour is not only directed towards relatively anonymous organizations, since we treat our work colleagues and families in the same way. It is common for this pattern of behaviour to start early, during our school days. So, when someone comments on our behaviour it is likely to focus on something detrimental that we would rather not hear about. Observers often tell us that they feel 'silly' praising the positive aspects of their colleagues' safety behaviour and feel uncomfortable about talking about unsafe behaviours because they perceive it to be a form of criticism. In our experience, the observer will need time to make this change and it will only happen when employees understand and believe in the behaviour-based approach. Whilst this is an issue that can be addressed with some success during observer training, we are aware that the observers are happier to rely on more visual forms of performance feedback, rather than verbal feedback in the early stages of their duties.

This has led to the development of methods of visual feedback. One organization had sets of photographs illustrating both safe and unsafe behaviours for each of the checklist items. These were displayed around the enlarged copy of the checklist. At the end of each week of observations, photographs of the three most safe and the three most unsafe behaviours observed were posted beside the feedback chart, with the captions 'Well done' and 'Need to do better', respectively. Other companies have produced a 'talking points' document, which serves the same purpose and simply highlights checklist items in terms of 'Well done' or 'Need to do better', or 'Need to focus'. This is also posted beside the feedback chart. (Figure 4.6 gives an example of this visual form of feedback.) It is usual for

Safety Behaviour Talking Points

- High levels of safety
  behaviour observed:

- Low levels of safety
  performance observed:

- . . . .

- . . . .

- . . . .

- . . . .

- . . . .

- . . . .

- well done!

- need to focus!

Figure 4.6 *Visual method of giving feedback in the workplace: safety behaviour 'talking points', usually produced weekly*

the observer to provide the immediate supervisor with a copy of this form for use in team briefing sessions or meetings where the barriers to safety performance can be discussed. Thus, if a certain checklist item continually appears as one of the top three unsafe behaviours, it can be addressed at both shop-floor and management level in order to find some solution to the problem. We have personally noted changes made to both systems and practice as a result of these discussions. Ultimately a consistently observed unsafe action is rendered safe.

Some observers prefer to provide daily safety performance feedback in a visual form which shows safety performance as 'percentage safe', 'percentage unsafe' and 'percentage not seen'. Usually these forms also describe one item which was seen as unsafe and in need of attention during the next shift. Figure 4.7 illustrates one example of this type of daily

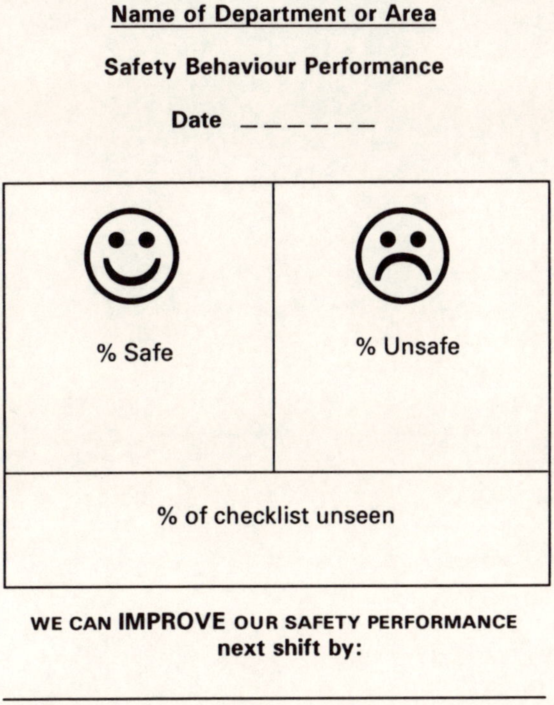

Name of Department or Area

Safety Behaviour Performance

Date _ _ _ _ _ _

WE CAN **IMPROVE** OUR SAFETY PERFORMANCE
next shift by:

Figure 4.7 *Visual method of giving feedback in the workplace: a daily safety performance chart*

feedback chart. It is usually on a laminated board, so the observer can wipe it clean and quickly prepare an update at the end of each shift.

Conclusion

These are just a few of the developments and changes made to our original approach to continuous safety improvement. Each organization finds new ways to make the programme its own and the majority of these ideas are generated by the observers themselves. We are fortunate to be able to share these ideas with other organizations and to continue to learn ourselves. With the right support and encouragement we have seen the level of observer confidence grow, with the result that both the observers, the programme and the organization all benefit. As the case study experience described above shows, the potential payoff can be high and rewarding.

The next chapter provides a guideline for the implementation of the behaviour-based approach to continuous safety improvement.

5

APPLYING THE BEHAVIOURAL
APPROACH TO SAFETY

As described in the previous chapters, the behavioural approach to safety is based on clearly defining the required behaviours and setting up a system to observe and record when these occur. The mechanism for doing this is to establish a checklist of safe behaviours, which is used for periodic random checks on activities in that part of the organization where the intervention is taking place. This information is then fed back to the participants, who use it to assess their progress towards safety goals and also, periodically, to revise and set new goals. Designing the checklists and carrying out the observations is done entirely by participants in the work activity in question. Checklists and the results of observations are the property of the group and need not be disclosed to anyone else. It is stressed that they may not be used for any sort of disciplinary action.

In outline the procedure is as follows:

1. Safe working behaviours are defined and a checklist of about 20 such safe practices is set up.
2. Volunteers from the work group involved in the study make random checks using the list to establish the present level of safe and unsafe working practices (the baseline).
3. This is shown to the group, who review their performance and set goals or targets for improvement.
4. Measurements continue for a further period of from 12 to 16 weeks, with continuous weekly feedback to the group. During this time safety behaviour invariably improves towards, or can even exceed, the target set.
5. At the end of this period the checklists and the goals are reviewed and updated and the cycle starts again. Usually at this stage new observers take over so that, over time, all members of the work group take part in the process.

In this chapter the complete process of setting up and managing such a behaviourally based safety intervention will be described. The sequence of the first two or three stages may vary according to the circumstances of the intervention, but it is important for a successful outcome they are all dealt with at some time.

Initial Briefings

Before the start of the behavioural programme it is usually necessary to undertake some 'awareness raising' activities. Raising the profile of safety issues helps the implementation of the behavioural programme itself. This awareness raising is accomplished in a variety of ways. A number of briefings are usually given to senior members of the organization (that is, those concerned with making the decision that the project should proceed), explaining the theory underlying the approach and outlining the stages of the project. When a decision has been made to go ahead, a letter of intent, signed by the chief executive (or the senior manager in charge), is sent to all employees. This states the aims and objectives of the programme and signals that there is a high level of commitment to it, at the most senior levels of the organization. This can be backed up with notices on bulletin boards and items in company magazines and newsletters. Only after these activities have been completed should the move be made into the next stages of the intervention.

Measuring the Safety Culture

At an early stage of the safety programme we find it useful to arrive at some indication of the existing levels of safe behaviour and of attitudes to safety, so that a suitable starting point can be defined. This is what we mean by 'measuring the safety culture' of the organization. The first step in this process is to interview a sample of employees, on the site in question, to determine their attitudes to safety and their perceptions of the organization's concern for the well-being of its employees. Even where a company does put considerable effort into establishing a safe working environment, there is often a widespread belief that management is more concerned with production than with safety. It is useful to know about these attitudes as it may give some indication of likely resistance to be overcome. The size of the sample to be interviewed depends, to a large extent, on the context, but should be enough to constitute a representative sample of the group who are to be involved in the project (say 10 per cent). The technique used is an in-depth, semi-structured interview. It is semi-structured to leave room for the interviewees to raise issues of concern to them. The interviewer should remain alert to possible leads into such concerns as they can often provide unexpected but vital information for the coming stages. The interview, which usually lasts anything from half an hour to an hour, focuses on the individual's attitudes to safety, accidents he or she has witnessed or been involved in, and potential hazards they see in the environment. As well as providing information for the coming stages, the interview sessions also allow the consultants to explain the nature of the forthcoming programme. This information tends to filter through to the rest of the workforce by 'word of mouth', and helps the process of disseminating information about the programme.

Based on the findings of the interviews, a questionnaire is designed to measure the safety climate of the organization. This is distributed to all members of the organization if the numbers are relatively small, or to a representative sample in larger organizations. The purpose of this is, of course, to provide more data so as to check that the findings from the interviews do apply to the larger group. Finally, a report is prepared feeding back the overall findings to the organization. This information is also useful, at a later stage of the project, when the initial checklists are being developed. In order to ensure that the information obtained, and included in the report, is as accurate as possible, a guarantee is given to all participants that their responses are confidential and no individual will be identified in any way.

The Initial Project

In all but very small organizations it is necessary to make a decision as to where to start. The usual process is to commence the intervention with a pilot project and, when this is running well, to extend out to other sections or departments. This makes it possible to concentrate resources in the initial stages, thus increasing the chances of success, and providing a demonstration of the technique for other parts of the organization. It is obviously necessary to select the area for the pilot scheme with some care, as it is desirable to start with a clear success – success, in this case, being measured by a significant drop in accidents within two or three months. Useful criteria for selecting the area are:

- Accident statistics should be available.
- It should be possible to demonstrate some improvement in safety – that is, the area should have a good but not perfect accident record.
- There should be a relatively stable workforce.
- Trade union and safety representatives should be active and involved.
- The areas to be included should be coherent and definable.

Staff Briefings

After the decision has been made concerning in which part of the plant the initial intervention is to be made, it is essential that all those who are going to be involved are thoroughly briefed. These briefings will be not unlike, but with more detail, those given to top management, during the original decision making process, when the project was initiated. It is important that everyone who is going to be affected, in any way by the intervention, takes part in these briefings. This will include not only those who are going to be doing the observing and being observed, but also, and most importantly, their supervisors and managers and anyone else involved with the department. During each session, which takes about one hour, an outline of the theory and a brief description of the process of the behavioural approach to safety is given. We find that this is desirable so that everyone understands the approach and why we are doing it. It is stressed that all observations are carried out by members of the group being observed and all results remain their property, and that we shall shortly be asking for volunteers to become the first set of observers. It is also made clear that no disciplinary action will result from anything on the checklists, which are confidential and need not be shown to anyone outside the group. The involvement and support of supervisors is absolutely vital, and it is essential that they fully understand and accept what is happening. For this reason they are usually given separate, and more detailed, briefings (see below).

Analysis of Safety Records and Initial Checklists

As explained in Chapter 3, for the behavioural approach to work, the behaviour to be influenced must be both observable and capable of precise specification. This is because the system depends on an observer, using a checklist, being able to assess unambiguously, whether a particular behaviour is, or is not, taking place. In the example described in Chapter 3 (the intervention by Zohar and Fussfeld (1981) in the Israeli textile mill) the behaviour easily satisfied both criteria. Ear defender usage is both an observable and a precise behaviour. It is very clear to an observer whether an individual is, or is not, wearing them. With a programme where the objective is to improve safety in general, the situation is not as clear cut. First, the specific behaviours that are to be encouraged have to be identified. The identification of these behaviours is made largely from two sources. First, an analysis is carried out of all accident records for the past two (or even three) years. The second source of information is the interviews described above. As we now know (see Chapter 1), many minor accidents, and near misses, go unreported, and these interviews help to discover the behaviours which underlie such incidents. While these may not be apparent from the formal accident statistics, they will, nevertheless, give important clues to unsafe behaviours. Other ways of identifying safe behaviours include the use of 'focus groups', brainstorming techniques, critical incident/'near miss' investigations, and interviewing those individuals who have been involved in a number of accidents.

From this information a 'checklist' is constructed for each department involved, comprising those behaviours which, if not done correctly, are most commonly associated with accidents. Examples of such behaviours might include 'goggles worn when using nail gun', 'cut resistant gloves worn when cutting' and 'long gloves must be used when recovering viscose from acid'. These checklists, of between 12 and 20 items, are further refined during the observer training process described below. (An example of a completed checklist is given in Figure 5.1.) There are two columns headed Observation 1 and Observation 2 because, ideally, observations should be made twice during each day or shift.

Training for Supervisors and Line Managers

For a successful intervention it is very important to ensure the involvement of the line managers and supervisors of the relevant groups. Special briefings are held during which it is stressed that their support is vital to the success of the project, and that it will in no way undermine their role as managers. Emphasis is placed on the problems and difficulties that may occur and what they can do to help overcome them. In general, the aim is to enlist their support and cooperation in the project. It is necessary for managers to support the observers and help the process to run smoothly without taking it over. Many managers find this quite a difficult balance to maintain. The problem is that if the manager is too controlling, he/she keeps the observers in a position of dependence and they do not feel empowered to take over the running of the scheme. If he or she stands back and takes little or no part, the observers feel unsupported. For this reason, the manager briefings include not only the theory of behaviour modification and the background to the behavioural approach to safety, but also information on concepts such as authority, dependence and resistance to change. These concepts and their significance are discussed in detail in Chapter 6.

The Role of the Observer

We now come to one of the most important aspects of the behavioural approach to safety – that is, the role of the observers. Their role is to provide measures of workplace safety by regular, systematic and random sampling of the activities taking place in the defined area. This is done using a checklist, which is developed by the observers themselves, based on the preliminary list provided by the consultants. It is stressed throughout that ownership of the project (which includes the checklists and results of the observations) lies with the participants. The observers are participants in the normal work of the department or section, who simply take a few minutes at random, but convenient, times to walk round the work area checking which activities on the checklist are being done safely. Observers are also responsible for scoring the checklists and providing feedback to the rest of their group or department. This is usually done by marking up charts placed strategically around the workplace. In some instances the information is made available on computer terminals in locations where these are available to the workforce.

The observers are absolutely vital to the success of the process. They are the main driving force in the behaviour-based approach. Success is largely determined by their motivation and commitment. It is essential that they are volunteers who thoroughly understand the process and are concerned about safety. In our experience there are usually enough of such people in any organization to start the process off. Others may, initially, be more

	Observation 1			Observation 2		
	100% SAFE	UNSAFE	NOT SEEN	100% SAFE	UNSAFE	NOT SEEN
1 Floor/walkways/steps/stairs to be free of tripping/slipping hazards	✓					
2 Chemical spillage to be treated immediately	✓					
3 Machines to be kept clear of waste		2				
4 Guards to be in place when machines are running	✓					
5 Safety footwear to be worn at all times	✓					
6 Ear protection to be worn in designated areas		1				
7 Eye protection to be worn when handling solvents			✓			
8 Mask and gauntlets to be worn when handling caustic		2	✓			
9 Gloves to be worn when handling waste		2				
10 Gloves to be worn when pushing reels	✓					
11 Dust masks to be worn when mixing, cleaning burning filters and baler filters			✓			

	Observation 1			Observation 2		
	100% SAFE	UNSAFE	NOT SEEN	100% SAFE	UNSAFE	NOT SEEN
12 Eye wash stations to be accessible; seals not broken	✓					
13 Hoses to be coiled/stored when not in use	✓					
14 Stanley knife blades to be retracted when not in use	✓					
15 Solvents to be stored in designated areas	✓					
16 Fork truck to be used in accordance with mill procedures	✓					
17 Trucks to be parked in designated areas when not in use		✓				
18 Fire exits to be kept clear	✓					
19 Fire extinguishers in place/seals not broken	✓					
20 Lifting equipment – blocks/slings/tackle/chains – free of damage	✓					
TOTALS	13	6	3			

Figure 5.1 *Example of a completed critical safety behaviour checklist*

sceptical, but as they see the improvements produced they become actively involved, so that after a while, everyone becomes committed to the success of the project.

The seven step approach to observation

The role of the observer can be summed up in the following seven step approach to observation. This is a framework, derived from Krause and Finley (1993), which we have found helpful in the training of observers and which we present to them on the training programme.

1 BE ACTIVE AND VISIBLE AS AN OBSERVER
Observations should be made from where things are happening. The observer should move around the work area, not try to observe from one vantage point. It is also important that people in the department or work area see the observer completing the checklist as this provides a reminder to think about their behaviour in terms of 'safe' and 'unsafe' acts, and helps to reinforce the fact that this is not a clandestine or underhand activity.

2 INTRODUCE YOURSELF
Colleagues should know exactly what the observer is doing. He/she should remind them that he/she is concerned with safety improvement. They should continue working as normal and results will be explained when the observation is finished. Everyone must know what is being looked at and why. It is essential to be open – the observer is not a spy. People should be assured that no names are recorded and no disciplinary action will result from observation. The observer is a champion of the safety improvement process.

3 LOOK AT PEOPLE AND LOOK AT WHAT IS GOING ON
The function of the observer is to check whether things are being done safely under safe conditions – particularly the items on the checklist. He/she should look to see if people are doing things correctly and safely. Is there potential for injury from objects in the environment or the general conditions? It is also necessary to look for 'footprints' of unsafe behaviour. Unsafe behaviour can often be directly observed, for example, improper use of tools or not wearing protective clothing, on other occasions the observation is indirect, for example, someone fails to replace a guard after working on a machine. In the first case it is possible to see the actual action, in the second instance the 'footprint' of an unsafe act can be observed.

4 LOOK FOR POTENTIAL INJURY
The observer must stay aware of the total situation and be alert to potential accidents. If someone is about to get hurt, it is important to stop the accident from happening. If potential hazards are seen, which require action (for example rubbish in walkways), the action should be taken. The checklist can

be completed afterwards. Preventing accidents is more important than a perfectly completed checklist.

5 USE THE CHECKLIST

The checklist should be used systematically, recording what is safe, unsafe or unseen. At first it will probably be best to start at the top of the list and go through it item by item. With practice and when familiar with the list the process becomes automatic and items can be recognized and ticked off as they occur, anywhere on the list. Care should be taken, however, not become too hurried and miss things due to over-familiarity.

6 CALCULATE THE SAFETY PERFORMANCE MEASURE

Observers complete and score the checklist and then calculate the safety performance measure. The procedure for doing this is explained in the following section.

7 GIVE FEEDBACK

An important part of the role of the observer is also to give feedback to the group. This can be on a one-to-one basis concerning any particular unsafe behaviour seen, or to the group as a whole, concerning more general matters relating to their performance on the checklist items. The observer is also responsible for more formal feedback via the feedback charts. Examples of these, with explanations, are given below.

The observer training programme

An outline of the programme for a typical observer training course, which usually lasts one and a half days, is given in Appendix A. While there may be slight variations in particular circumstances, the objectives of the programme are as follows:

- To develop an understanding of the theory and practical application of the behaviour-based approach to continuous safety improvement.
- To learn the skills of producing and using a critical safety behaviour checklist.
- To understand the practical issues associated with the observation of safety-related behaviour, goal/target-setting and providing performance feedback.
- To understand, recognize and manage resistance to change.
- To provide an opportunity to practise the skills of observing and measuring safety performance.

Training on each of the above areas is given during the programme, each of which is discussed in the relevant sections below.

The Checklist and the Safety Performance Measure

Scoring the checklist

As indicated above, the observer should walk round the designated area at convenient, but as far as possible, random intervals and, using the checklist, score the number of safe and unsafe acts as described below. It is important to score only what is seen. The observer should not try to interpret behaviours or make assumptions. Only what is actually seen should be recorded. The method of recording is as follows.

The observer puts a score of '1' in the '100% SAFE' column to indicate safe behaviour. For example, if all workers are seen wearing safety footwear in the area where it is specified that they should be worn, a score of '1' is put in the 100% SAFE column on the checklist (see item 5 in Figure 5.1). This indicates that this item has been observed to be 100% safe on this occasion. All workers present and seen must be behaving in the specified safe manner for a score of '1' to be given for any item in this column. If, however, any workers are not wearing safety footwear in the specified zone, a zero ('0') is entered in the SAFE column (or it can be left blank), and the number of people seen to be acting unsafely is entered in the UNSAFE column. In the example two workers were not wearing gloves when handling waste, so a score of '2' is entered in the UNSAFE behaviour column for item 9. In this case, obviously, the 100% safe column has a zero or blank. Thus there are only two possible scores that can be recorded in the safe behaviour column:

'1' if everyone is seen to be behaving in the specified safe way.

'0' (or blank) if at least one person is seen behaving unsafely.

The UNSAFE behaviour column can have scores ranging from '0' if all observed behaviour is safe, or from '1' to as many unsafe behaviours (i.e. individuals not behaving safely) as are observed.

The UNSEEN column is used to record any item of behaviour not seen on a particular observation round. In the example checklist, at the time of the observation, no one was seen handling solvents (item 7), so a score of '1' is entered in the UNSEEN column.

The safety performance measure

When the checklist has been completed the safety performance measure can be calculated. This is a weighted scoring system, which is very sensitive to even slight changes in the frequency of unsafe behaviours. It is, simply, the number of safe behaviours seen, expressed as a percentage of the total number of safe and unsafe behaviours observed. To calculate this ratio, the SAFE and UNSAFE columns are totalled and the following formula is applied:

$$\frac{\text{Total number of safe behaviours}}{\text{Total number of observations (total of safe + unsafe behaviours)}} \times 100$$

This measure is used to provide feedback to the group on how their safety performance is changing, usually of course, for the better. The exact procedure for this is described in the section on the feedback process, below.

The percentage unseen

It is also useful to calculate a percentage unseen figure. This is the total number of unseen behaviours on any checklist expressed as a percentage of the total number of items on the checklist. Expressed as a formula, this is:

$$\frac{\text{Number of 'unseen' behaviours}}{\text{Total number of items on the checklist}} \times 100$$

This figure provides an indication of the significance of the safety performance measure. Obviously, if the percentage unseen is very high, less reliance can be placed on the safety performance measure than when the percentage unseen is very low.

Establishing the Baseline and the Baseline Review Meeting

Before the intervention proper begins it is necessary to establish the 'baseline', that is to define the existing level of safe behaviours, against which improvements can be measured. To do this, observations are taken and the safety performance measure is calculated for a period of some four weeks before any feedback is given by the observer. This baseline safety performance measure is the average safe behaviour for the complete four-week period. At this point, when the baseline measure has been established, we would normally hold a review meeting with the observers. This has a number of purposes:

- To check and, if necessary, make minor amendments to the checklists (but make no major changes at this stage).
- To deal with any problems and difficulties encountered by the observers.
- To list and remind management of any safety improvement work that needs to be done in the plant.
- To identify and resolve possible barriers to the success of the intervention.

Goal-setting

When the baseline is established the observer is responsible for convening a 'goal-setting meeting', where the group is given feedback about its level of safe behaviour and then defines a goal for improvement, that is, a target for an increase in the overall index of safe behaviour. This should be set at a level that indicates a real improvement, but one that the group also feels is attainable. It is important that the goal is accepted by everyone: goals that have been set participatively are far more likely to be achieved than those that have been imposed. So to this end the observer should hold discussions with the group he or she has been observing, showing them the current level of performance (i.e. the baseline measure) and agreeing the new goal. Ideally, this is done at a group meeting where everyone is present, but as this is not always practicable, sometimes the observer will hold discussions with individuals or sub-groups and arrive at a consensus in this way. However this process is carried out, it is essential that everyone agrees with, and is committed to, the goal set. This target should be marked in a distinctive way on the feedback chart (see below).

The Intervention Period and the Feedback Process

After the baseline and goal for improvement is established, the observer continues with observations for a period of between 12 and 16 weeks (different organizations use different intervention periods) and calculates the average percentage safe behaviour, week by week, and provides regular feedback to the group. This is done in two ways – visually and verbally. Notice boards are placed in strategic positions around the site, where they can be seen and consulted by everyone. These display a chart (an example is given in Figure 4.1), on which the observer marks the goal set by the group and then, week by week, plots the actual percentage of safe behaviours observed. If all is going well, this plot will approach and, possibly, eventually exceed the set goal level. The observers should also give informal verbal feedback to individuals and the group. The value of this is that it is possible to focus in on any particular behaviours that may be affecting the overall safety performance. Thus if one, or more, individuals are consistently behaving unsafely on a particular item on the checklist, this can be brought to their attention and, hopefully, corrected.

This is usually the most difficult aspect of the programme for the observers. They need to be encouraged as much as possible and reminded that it is important to give positive feedback (in the form of praise) in addition to explaining to colleagues how they can improve their safety performance measures in the future. Effective feedback is a skill that takes time and practice to develop. Observers must be reminded continually that giving information on performance is not criticism. The only objective is for the workgroup, or shift, to achieve a safer work environment for the benefit of everyone. The necessary climate of openness and trust (see Chapter 6) does not emerge overnight, and so observers must not become discouraged in the early days and need to be reminded that progress will be slow. The support of supervisors and line managers, at this stage, is crucial, and management at all levels must show high commitment to the programme.

Other contexts in which feedback can be provided are during team briefings, where safety issues are being considered, and at specific safety meetings. Care must be taken in these contexts to maintain the guaranteed confidentiality, in that information concerning any identifiable individual should not be made available. There are also a number of other charts which can be used to give more general feedback, to the group as a whole, about particular items on the checklist. Example of these are given in Chapter 4 (see Figures 4.6 and 4.7).

Developing Continuous Safety Improvement

After the end of the first 12–16-week cycle it is usual to change observers, but, in any case, at this point there should be a review meeting where both the checklist and the goal is reviewed. By this time it may have become apparent that there are other items which should be on the list while some existing items may have become permanently 100 per cent safe and so can be dropped. If the existing goal has been met or exceeded the group should also set a new one. The observations and feedback then proceed for a further 12 /16-week period, followed by another review and goal-setting meeting. These cycles continue with each member of the group taking their turn as observer. In this way observation, feedback and continuous safety improvement becomes part of the normal ongoing activity of the work group. It is our experience in many organizations that this activity quickly becomes accepted and taken over by the employees themselves and provides an example of true empowerment, in that groups of individuals are now taking full responsibility for an important aspect of their working life.

6

MANAGING THE CHANGE

Implementing the behavioural approach to safety involves participants in making a significant change in the way they work – a change not in working methods or the detail of their job, but an add-on in terms of making the checklist observations and being responsible for goal-setting and feedback. This is an activity with which operators will be unfamiliar and they may find it rather daunting, involving, as it does, new and unfamiliar responsibilities, and ones which they fear may bring them into conflict with their co-workers. This may, and often does, give rise to some degree of resistance to the programme. This resistance may also be linked with other concerns about safety and beliefs about management attitudes to creating safer working conditions. We always start observer training programmes with a 'hopes and fears' session to get these beliefs into the open. This is so that we can, or so we hope, allay the 'fears' and encourage the 'hopes'. Some of the commonly expressed fears are as follows:

- It will make life more difficult or more complicated.
- This is just another management fad that will be abandoned in a few months.
- The information will be used by management to make scapegoats when accidents happen.
- Observers will become management 'spies'.
- Concern at the possibility of being asked to become an observer, or take some other role, often coupled with fear of having to take more responsibility.

It has to be admitted that sometimes there is justification for these responses. It is not surprising that participants do occasionally feel that this is this just another management fad which is going to be forgotten in a little while, when the next fad comes along. Unfortunately, the changing fashions and 'flavours of the month' in management practices do provide some foundation for this cynicism. Another common response is that 'management is not really concerned with safety – all this will be forgotten when there is pressure on for production'. Although most managers would, of course, categorically deny that they ever sacrifice safety for production (and do genuinely mean this), the reality is that managers, being human, tend to concentrate on whatever is the most pressing problem at the time. If the problem is meeting deadlines, other issues will become secondary, and the manager may tend to forget about safety for the time being. (See discussion of reinforcement in Chapter 3.) However, we believe that managers are, on the whole, sincere in their concern to create a safe working environment. For us, the behavioural approach is not just another fad. It is the application of a well-tried and well-established technique to improve safe behaviour. However, when such fears are expressed, they do need to be taken seriously and responded to in an appropriate way.

But, perhaps, by far the most important factor in creating resistance to the behavioural approach is that it requires a very considerable change in the way people work. Individuals are now being given autonomy and responsibility for an aspect of their work life that they may never have had before. As indicated in Chapter 4 and discussed more fully in Chapter 7, the behavioural approach can mark the start of a process of real empowerment. Many employees in large companies have very little (or no) experience of taking any responsibility during their work lives, and it can seem quite frightening when they are suddenly asked to do this. The explanation for this lies in the way people in organizations have traditionally been managed and how this process is changing. In the next part of this chapter we present a number of theories concerned with organizational and individual change. These are all theories which we draw on during training and implementation of the behaviour-based approach to safety management and which, we find, help to give participants an insight into the process and helps them to adjust to the change. We will present the theories first, with only brief reference to safety, and then explain their implications. We ask readers to bear with us if the next section seems a little theoretical. We will return to practical issues at the end of the chapter.

Theories of Organizational Change

Three phases of organizational development

The first theory defines three types of organization and suggests that these may be a developmental sequence. Lievegoed (1973) has suggested that organizations tend to develop through three well-defined phases. These phases he calls 'pioneer', 'differentiation' and 'integration'. Each is characterized by its own structure, styles of management and culture, which have a significant effect on the way people work within them. Each phase also has its own particular problems and, according to the theory, it is the attempt to resolve these problems that pushes the organization into the next phase of development. Of course, not all organizations start in the pioneer phase, some can start as differentiated, or even integrated, organizations, alternatively others can skip the differentiation phase and go from pioneer to integration. This is simply suggested as a general trend. In fact, for our purposes it does not matter whether the developmental theory is correct, since all three types of organization do exist, and there are significant differences in the way people are managed in each of them. Organizations can also be in a transitional stage, showing some elements of two phases. For example, different parts of one organization can be more, or less, integrated or differentiated than other parts. The three types of organization are described below.

PIONEER ORGANIZATIONS These are typified by small entrepreneurial companies. They are run by the entrepreneur (or a small group close to the entrepreneur) who tends to dominate the whole organization. He or she makes most of the decisions and can communicate them directly to the people involved in carrying them out, and can monitor progress more or less directly. Because the organization is small, most people are in fairly direct contact with the entrepreneur. Communication is, therefore, fast and reasonably efficient. The organization can adapt and change quickly. If the entrepreneur says so – it happens. Effective change depends on the entrepreneur's ability to recognize what change is needed and understand how to carry it out. Employees are usually quite involved and willing to adapt and take initiatives, provided they are sure they have the backing of the entrepreneur. There is not too much problem with resistance to change. If there is resistance this tends to be overcome with direct authority. However, if the company is successful it grows larger and it becomes impossible for the owner/manager (or group) to continue to control all aspects of its activities, mainly because rapid communication and decision making becomes difficult, if not impossible. To cope with this situation specialist departments and sections are set up to deal with specific aspects of the company's operations, that is functions such as accounting, sales, personnel and so on. Thus the organization moves towards the next phase, that is, differentiation.

DIFFERENTIATED ORGANIZATIONS The move towards differentiation is an attempt to solve the problems of communication and control created by the growth of the organization and the associated employment of specialists. The thinking behind this form of organization is strongly influenced by the traditional writers on management and is based on a machine metaphor. That is, the organization is thought of as rather like a machine, which once set up and running will keep going more or less indefinitely, with the occasional maintenance and replacement of worn out parts. People are thus treated like machine components whose role is simply to carry out the tasks assigned to them. This is, still, by far the commonest structure within large organizations. It is characterized by specialization and separation on a number of dimensions: for example, by function, by product, by location and by hierarchy. With large numbers of employees, communication and decision making needs to be controlled in some way. This is often done by creating rules and procedures that will eliminate the need for personal guidance from the top for routine matters. Often there is a specialist department whose sole function is to keep the rulebooks and system up to date, and disseminate them throughout the organization. This has the effect of reducing the level of discretion of organization members; particularly those at the bottom of the hierarchy. Mechanistic systems for managing people come into operation. Jobs are carefully specified and filled by systematic selection procedures. Disciplinary procedures are codified. The hierarchy is clearly specified and there are formal lines through which people report. What we are describing is, of course, the traditional 'tree' organizational structure, with the chief executive at the top, delegating control through a formal system of successive layers of management.

Such organizations are characterized by what Harrison (1987) called a *role culture*. As with pioneer organizations, power is still very much in the hands of top management. However, it is not exercised at personal whim, but by reference to formal systems and procedures. The high degree of formalization means that individuals are protected against arbitrary power, as long as they stay within the rules. There are, unfortunately, a number of disadvantages to this type of organization, as everyone who works within one knows. The high degree of formalization and the large number of rules tends to create alienation. This is because such organizations have a depersonalizing effect on individuals, who often feel that they have little influence and are under-utilized. This can apply to even quite senior members. There is, also, often dysfunctional conflict between department and sections. Decision making is slow and the organization finds it difficult to respond to change. Because they tend not to feel valued in this culture, individuals tend to comply mechanistically with the rules and requirements of the organization, but withdraw their commitment and initiative and become resistant to change. Thus, implementing the behavioural approach to safety can present some difficulties in this context. It is necessary to find a way of getting people involved and willing to take responsibility.

INTEGRATED ORGANIZATIONS The reality is, that because of their complexity, most large organizations do need specialists and a high degree of diversification, but they also need some mechanism to overcome the problems of a high degree of differentiation, as outlined above. In other words, they need some form of *integration*. There are a range of methods by which an organization may achieve this. The particular method chosen, and the end result, will be specific to the organization, but will share certain common characteristics. There will, almost certainly, be more adaptability in styles of leadership. There may be greater emphasis on teamwork and team performance, rather than that of individuals. Decision making will be more decentralized with the use of more participative techniques. This form of organization is based on an organic metaphor, rather than a mechanistic one. It is not seen as a static machine, but rather as a growing, living organism. In systems terms it will be seen as an open system, where development and change are viewed as inevitable and natural.

Unlike organizations in the differentiation phase, which tend towards a uniform, all-pervading culture, organizations in the integration phase show a variety of cultures, depending on their function and form (see, for example, Harrison's (1987) *support* and *achievement* cultures). There may even be different cultures in different parts of the organization. Whatever the culture, the emphasis will be on the value of the individual and the contribution he or she can make, while also acknowledging the desire people have to achieve their own personal needs and goals within the organization. There will need to be systems for achieving compatibility between these personal goals and those at an organizational level, but with an emphasis on action and initiative. As with pioneer organizations, people working in this type of environment develop high commitment to the company, but in this case also feel free to trust their own judgement and take personal initiatives. Change, if it is seen to be useful and desirable, will be readily accepted. Employees will even suggest and initiate change.

The implication of Lievegoed's theory is that it may be very much more difficult to initiate change and get individuals to accept responsibility in some organizations than others, and that a different approach to initiating change may be required in different types of organization. Essentially, the problem is how to get individuals who have had very little or, sometimes, no experience of taking responsibility during their work life, to start taking control of their own, and fellow workers' safety behaviour. Cox and Makin (1994) have suggested a theory which helps with this problem.

From authority to interdependence

This theory centres on the concept of *dependence*. Whenever authority is being used (as in role cultures), the individual on the receiving end of this authority is inevitably pushed into a position of dependence. That is to say, that because they are always told what to do, and do not have to make decisions for themselves, they eventually become dependent on the authority figure and will not know how to act unless told what to do. Because they have no experience of initiating action, and subsequently being responsible for it, they will be unable to do this, even in situations that, ideally, would require it. Because they have habitually been told what to do, they have become relatively passive during their work life. The longer they have been in this situation the more complete will be their dependence. Most of us do, in fact, get a great deal of experience of dependence. As young children we are dependent on our parents – when very young, literally for our survival. At school, college and in our early working life, we also spend much of our time in a situation where others make the decisions and we do as we are told. For many people this continues throughout their working lives. In fact, given this background, it is surprising that we do not get more resistance than we do to the changes involved in setting up initiatives such as the behaviour-based approach to safety. However, where resistance is encountered, how does this concept of dependence help us to cope with it? To explain this we need to define some other possible relationships which are connected with authority and dependence. These are shown in Figure 6.1 below.

COUNTER-DEPENDENCE This arises where the dependent individual wishes to escape from being controlled, but the authority figure will not give up control. In this case, all that the dependent person can do is 'fight' back. This will often take the form of some act of rebellion. Whatever the action taken, this is the way it will almost certainly be seen by the authority. 'Wild cat' strikes are good examples of counter-dependence in organizations. Often, the workers concerned feel that this is the only way they can make their feelings known. Strangely enough, a counter-dependent reaction can also arise in what is almost the opposite situation, such as when the authority attempts to hand over control before the dependent individual is ready to take responsibility and make their own decisions. In this case the 'fight' is to push the authority back into control. Thus, counter-dependence is really about establishing who is in control and of what.

MUTUAL DEPENDENCE This is a more cooperative situation. Control is shared by mutual agreement, which may, or may not, have been openly negotiated. Usually, in organizations the mutually dependent roles have evolved over time, without anyone being too consciously aware of the inter-relationship or its implications. However it arises, it always takes the form of each party controlling their own areas of activity, but in a way that is mutually supportive of each other, with neither party taking any responsibility for the other's activities. It is usually quite a powerful relationship, but has the weakness that if one half of the partnership fails in its part, the other half will not pick up their activities. This means that, at best, jobs will be left half done or, at worst it will cause the collapse of the whole partnership. It is quite a common relationship in the management of professional organizations (accountants managing other accountants or lawyers managing lawyers), where manager and subordinate have equal respect for each other's skills and abilities, and each is confident about their own role, but it does also occur in manufacturing and other types of organization. In mutual dependence both manager and subordinate know their jobs and get on with them, dovetailing together without the use of authority.

INTER-DEPENDENCE Inter-dependence may, on the surface, look rather similar to mutual dependence. The partners may have a tendency to concentrate on their own areas of activity, but they will, if need arises, be capable of taking over each other's roles. Usually, there will be considerable flexibility and interchange of activities. There will be joint decision making and sharing of control. The relationship will be one of cooperating equals. It is usually a very rewarding and creative way to work, but can be difficult to maintain. Effort and commitment is needed to resolve conflicts and disagreements. In management terms this will be a truly participative system, each partner concentrating on their own job, but able to take over, or assist with, the other's if the occasion demands.

INDEPENDENCE The final situation to be defined here is not strictly a relationship at all. In the context of this theory independence usually arises from an unsatisfactory resolution of counter-dependence, where the conflict is avoided by one party withdrawing from the relationship and proceeding to operate on their own. There is a place for independent working in organizations, but it is obviously better if this is as a result of a conscious decision that this is the best way to work in a particular situation, rather than as a result of an inability to resolve issues of authority and control.

MOVING FROM DEPENDENCE TO INTER-DEPENDENCE Fairly obviously, the implementation of the behavioural approach to safety (and indeed many other change initiatives) will only be effective in a situation of inter-dependence. If there is a high level of dependence on the part of participants, it will be very hard, if not impossible, to get them to take responsibility for checklists and observations. In fact, getting the balance right is quite difficult. Management must be involved and show interest and support where necessary, but must not take over, so that participants lose owner-ship, or withdraw too far, so that participants feel they have been abandoned.

Since, as indicated above, many organizations have authoritarian role cultures, which will have developed relatively high levels of dependence in their employees, the problem arises as to how it is possible to move people from dependence to inter-dependence. Unfortunately, experience teaches us that there is no way to move directly from one to the other. It seems that it is only possible to move via counter-dependence or mutual dependence, or quite often, a combination of both. This is indicated by the arrows in Figure 6.1.

Many managers will discover that, if in an attempt to move towards greater participation they attempt to give up control too quickly, sub-ordinates move straight into counter-dependence. This will usually involve arguments as to who is responsible for the activity or decision involved,

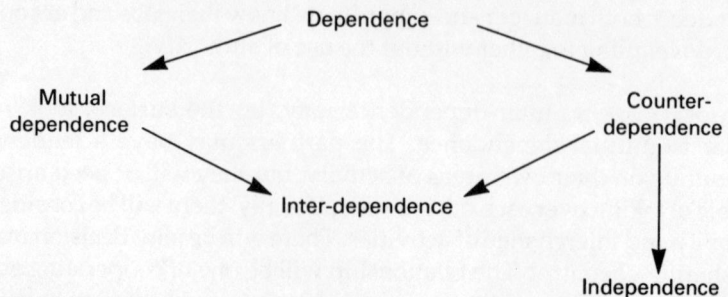

Figure 6.1 *Dependence relationships*

with claims that it is the managers' job to take charge. In other cases the response may simply be passive resistance; things simply do not get done. There is evidence that if the person in the authority position persists and refuses to take back control, then eventually the dependent partner(s) will start to assume responsibility. Within organizations this is obviously a fairly dangerous strategy. While the issue of control is being fought out, very little work may be getting done, and the situation can escalate out of control. There is also the danger that if the pressure becomes too much, people will move into independence, either by psychological withdrawal, or actually by leaving the organization.

A safer way to move is via mutual dependence. This involves negotiating to hand over control to the subordinate(s) until an equal sharing is achieved. A possible problem with this strategy is that mutual dependence is very comfortable. Because roles are clearly differentiated and work is shared, nobody feels over-controlled or over-responsible. It is also relatively easy to manage because it is clear who is supposed to do what. Consequently, there is a tendency to get stuck in this stage and not make the move on into interdependence. In reality, it is very hard to judge the rate at which control can be transferred, so there is always the risk of flipping over into counter-dependence. This means that most change processes will, to some extent, involve both routes – some pressure from management and consequent argument and some negotiation and smooth transfer of control

To give a little more insight into this theory, it is, perhaps, worth making an aside to point out that what is described above is, in fact, the process of growing up. We all start out in life dependent on our parents; in early infancy, for our very survival. The mutual dependence and counter-dependence stages are what are thought of as typical teenage behaviour. Although, in fact, they tend to last from around the age of three or four until the twenties (or in some cases, for life). What is going on, of course, is that the child is sorting out who is in control and of what. The wise parent will offer a gradual handover of responsibility but with safeguards ('You can stay out until midnight, if you phone in and let me know where you are'). Being fallible we will not always get it right, and the relationship will tip over into the arguments and tantrums of counter-dependence. If we persevere the relationship will eventually develop into that of two inter-dependent adults. If not handled properly, the child may retire into independence, either by psychological withdrawal, or more tragically, by simply leaving home and disappearing. It could be argued, by those who are psychoanalytically oriented, that the reason we have so much trouble with authority at work is because many of us have not fully resolved the problems of dependence which we have with our own parents.

Phases of development and dependence relationships

From a consideration of the two theories discussed so far, it becomes clear that large differentiated organizations involve a high degree of authority/dependence in the way that they are managed. Conversely, it is impossible for an organization to operate within the integration phase without there being a high degree of inter-dependence between managers and their subordinates. Thus the move from differentiation to integration involves the move from dependence to inter-dependence. In a differentiated organization members will assume that responsibility for all planning and decision making lies with 'the management'. This goes for all aspects of policy – long-term goals, coping with change and, importantly in this context, safety. When things go wrong it will be 'management's fault'. This is not, actually, an unreasonable view since, in this type of system, 'management' does make all the decisions. If a system is required where individuals accept responsibility for their own actions (as with the behavioural approach to safety), this requires a degree of inter-dependence. This raises the question of how, within a large, complex and authoritarian organization, the change from dependence to inter-dependence can be made. The paradox with which we have to cope is that the change has to be made in a way that is compatible with the system we are aiming to produce, and yet does not violate too greatly the existing beliefs and assumptions on both sides (i.e. managers and subordinates). In other words, one cannot change a differentiated role culture, towards inter-dependence, by making new rules. However, if counter-dependence is to be avoided some control must be exerted, and accepted, in the early stages of the process. The approach we are advocating in this book seems to achieve this. Just how and why will be discussed later in this chapter. Before doing this, we need to look at some other theories concerned with organizational change.

Sources of power or influence

As noted above, when attempting to create change in organizations there will often be a certain amount of resistance or, to use a mechanistic metaphor, inertia to be overcome. Lewin (1947), suggested many years ago that change is a three-stage process. The first stage is *unfreezing*. This involves shaking the organization members out of their current patterns of behaviour, or in other words, convincing them of the need for change. In the mechanistic metaphor this is the process of overcoming the inertia and starting a momentum towards change. When this is done it is possible to move to the second stage, of making the *change*. The final stage is *refreezing*, or establishing and stabilizing the new patterns of behaviour. For much organizational change this is a rather outdated model, since many organizations now seem to exist in a state of almost permanent change and the problem becomes not how to start change, but rather how to cope with almost continuous uncertainty. However, in the context of improving safe behaviour, which is a matter of breaking old unsafe habits and replacing them with new safe ones, this is a good description of the process.

To continue the metaphor, if we are to overcome inertia, some source of power is needed to start the system moving in the appropriate direction.

Managers have many sources of power at their disposal, in a social context also referred to as sources of influence. French and Raven (1959) suggest that there are five sources of power or influence available to managers (or consultants) seeking to create change. These are:

1 Reward power: This is the power which results from the ability to offer rewards for compliance. Additional financial rewards or productivity bonuses are often negotiated as an incentive to make changes in working methods, or payment can be made for the abandoning of outdated working practices. Rewards may not only be financial; time off, a shorter working week or better working conditions can all be seen as the use of reward power.

2 Coercive power: This refers to the use of punishment to influence others. It is not an effective method of influence in most organizations (see the section on punishment in Chapter 3), as individuals can usually find ways of avoiding the worst effects, and in most reasonably affluent societies people will simply not accept employment in a company where policies are too abrasive. It is used in organizations such as the military and prisons, where punishments are usually fairly swift and cannot be avoided, since once enrolled members cannot easily leave. It is also used by governments in the form of both the penal and the taxation system.

3 Legitimate power: This form of power is conferred on the individual by the rules of the organization. In most organizations, the more senior the manager, the more legitimate power he or she has. As managers rise up the hierarchy they usually acquire more authority and control over a wider range of people and activities. Civic authorities such as judges and police officers also have legitimate power. To be effective this form of power requires the acceptance of the rules, by everyone involved. If those on the receiving end do not accept the attempted control and therefore resist, the power figures frequently then resort to coercive power. However, in Western democracies, if resistance continues the power figures usually, ultimately give way and the rules are amended. Unpopular laws (and even governments) are often defeated in this way. The most recent example in the UK was, perhaps, the repealing of the community (or poll) tax. Even military dictatorships have been toppled because the dictator was not willing (or able) to wield sufficient coercive power to maintain his position. Similar, but usually less dramatic, events take place in organizations.

4 Referent power: This is the power possessed by charismatic individuals who inspire others to emulate them, usually by a process of identification. Their power resides in their personal characteristics and is not conferred from outside. Much social change is inspired by such people (for example Martin Luther King and the civil rights movement). Some management consultants and managers operate in this way.

5 Expert power: The possession of knowledge or expertise confers power on the holder, who can use this knowledge to influence others. This is

usually by a process of logical argument or persuasion. This is usually the most potent source of power for consultants, although they can, and do, combine this with any of the other sources. As noted above, they sometimes have referent power, but can also tap in by proxy, as it were, to the legitimate and reward power of their sponsor in the organization. This is, of course, why consultants make their entry as far up the organizational hierarchy as they can, where these sources of power are usually at their highest.

The effective manager or change agent needs to be aware of, and be able to use whatever sources of power he or she has available. These will vary with the context and with the individual. However, power or influence will not be effective unless the individual knows how to use it. Some knowledge of change strategies is also necessary.

Strategies for change

Chin and Benne (1976) have defined what they describe as three 'general strategies for effecting change in human systems'. For our purposes a 'human system' is an organization, although the strategies do also apply to wider social systems. The three strategies are:

1 Power–coercive strategies: These involve the use of coercive, resource or legitimate power to force individuals into changing or into accepting change. Requiring employees to sign new less beneficial contracts or accept redundancy, as has happened in many organizations, is an example of this strategy in use. In terms of safety it is the use of sanctions to enforce safety regulations, such as dismissal for breaking safety rules. This can be a legitimate sanction for such serious breaches of safety as bringing matches into a mine or petro-chemical plant. The problem is that while it certainly prevents that individual from offending again and may even provide some 'encouragement for others', it does very little to improve safety in general.

2 Empirical–rational strategies: This approach assumes that people are basically rational. Change can be effected, therefore, by showing that it is in the individual's own interest to change. This is achieved by the use of information and rational persuasion. The assumptions underlying this strategy are deeply embedded in traditional education and include a belief in the value of research and the dissemination of knowledge. The use of this strategy requires, primarily, expert power.

3 Normative–re-educative strategies: This group of strategies does not deny the rationality of human beings, but it places more emphasis on the belief that behaviour is largely determined by social and cultural norms. Individuals have a strong commitment to maintaining and conforming to these. Creating change using this strategy is, therefore, accomplished by changing norms. Referent power is highly effective in

this context. Culture change in organizations is largely attempted by this strategy. Since its culture is intimately connected to the prevailing norms within an organization this seems a reasonable approach. That it has been largely unsuccessful is probably due to a lack of change agents with sufficient referent power. This includes those working in safety improvement.

These strategies can be, and usually are, used in varying combinations. An interesting example of this, already mentioned in Chapter 4, is the British government's attempts in the early 1980s to get car drivers and passengers to wear seat belts. They started with a rational–empirical strategy. Considerable time, money and effort was spent trying to persuade people of the value of wearing seat belts in relation to improving their safety. Campaigns were run in newspapers and on television giving information on how seat belts save lives. There was some normative–re-educative back-up by using show business personalities (with referent power) using seat belts and extolling their virtues. All this had very little effect, so eventually the government moved to a power–coercive strategy by introducing fines for anyone not using a seat belt. On the day this law came into effect seat belt usage shot up to over 90 per cent. We suspect that this has now become the norm, and as discussed in Chapter 2, even if the law were repealed seat belt usage would remain at a high level.

While a power–coercive strategy works for governments, who have very high coercive power, it is a dangerous strategy in organizations. It is quite consistent with a differentiated role culture, but is likely to engender a high degree of counter-dependence and will, at best, lead only to compliance, probably, together with some degree of alienation. The other two strategies are quite compatible with both differentiated and integrated organizations. However, producing change by rational persuasion, based on data, is a very long-term process and may not, in any case, be successful. Similar comments can be made about the normative–re-educative strategies. Fortunately, a combination of the two can be highly successful, which is what is provided by the behaviour-based approach, as we explain below.

Theory into Practice

As already indicated, the most usual situation we find when starting a behaviour-based approach to safety programme is that of a differentiated organization with a role culture, and hence a high level of dependence. There is also, very often, some suspicion of the motives of management, or at least, a lack of belief that management has a serious interest in safety. (In an organization that has moved some way towards integration there will, by definition, already be a level of inter-dependence, hence the introduction of new responsibilities will not be so difficult, at best involving only discussion and agreement.) It is, therefore, necessary to find some way of unfreezing this initial situation and to select an appropriate strategy for change.

As external consultants our primary source of power is expert power, so we are mainly confined to using the rational–empirical approach, combined with the normative–re-educative. We can tap into the coercive, legitimate and reward power of management, but need to do this with caution as there is a danger of locking into the existing dependence and thus making harder any move towards inter-dependence and empowerment. In fact, we are careful to use only management's legitimate and reward power. Coercive power is most likely to generate counter-dependence. Hence, the first, unfreezing, stage consists of convincing management that this approach will work. This we do through a series of presentations where we explain the theory and show how it is applied in this context. We are now, also, able to provide evidence, from previous interventions, to show that the approach does work (rational–empirical approach). Once management is convinced and the decision is made to go ahead, we move to a mildly power–coercive approach. The chief executive sends out the letter of intent stating that the scheme is going to go ahead and explaining what will happen. To keep this within the use of legitimate power we stress that observers should be volunteers (no individual coercion) but that they will have the support of management and be given time to carry out the tasks involved.

We now move to a rational–empirical approach, as we start the observer training programmes. Our initial task is to persuade participants that the scheme will work, so that they are willing to try it, and then to provide them with the knowledge and skill to be effective observers. For this we are dependent on our expert power. As an aside, it is interesting to note, that while it is usually our intention eventually to hand over the running of the programme, for later phases, to internal staff, we have on occasion encountered some resistance to this from the participants, on the grounds that they felt they needed to be trained by psychologists, not safety officers. The reality is that in most cases we could quite easily train internal staff to take over from us. However, they are not seen by the members of the organization to have the necessary expertise. There are sometimes occasions when referent power can also be used. There may be a particularly charismatic manager who champions the process and simply by enthusiasm and example helps to move the process along. Similar power and influence sometimes resides with enthusiastic safety representatives or trade union officers who are concerned about safety. The support of such individuals is extremely beneficial to the success of a programme.

In effect, the approach we take facilitates the move from an authoritarian–dependent culture to a mutually dependent one, or in some cases even to inter-dependence. In the case of the organization described in the case study in Chapter 4, for example, where, after a very successful safety programme, the same approach was applied to improving quality (see Chapter 7), with equal success, the organization was effectively moved to much greater integration with considerable empowerment for members of the organization at all levels. This was only achieved, of course, because senior management became committed to the idea of allowing operators and other members of the organization more autonomy and realized that this was an effective way of bringing this about. The process was also facilitated by an enthusiastic champion in the person of the process improvement manager. It should also be acknowledged that other parallel and compatible initiatives were taking place in the organization at the same time. All of these helped to facilitate the change.

Relationships in Change Management

From what we have said above it is clear that, when introducing change, some resistance is almost inevitable. It is quite natural that people will be doubtful when faced with something new and challenging, particularly if it is seen as making more work or life harder for them in some way. Also they may not, at first, perceive the potential benefits. Even when a benefit such as safer working conditions may seem to be of obvious value, they may not believe that it is attainable. For effective introduction of change it is necessary to understand and be sympathetic towards the reasons for any such resistance. Remember, if people are raising doubts or asking difficult questions, they may have a valid point and may have seen a possible problem or, perhaps, a better way of doing something. There are a number of straightforward practical recommendations we can make which will help the introduction of such change.

- Be clear yourself about what is entailed in the change and communicate effectively what is involved. That is, what people are expected to do, how long it will take and what extra work is involved. Explain the benefits to be expected, i.e. a safer workplace.
- Remember that effective communication involves listening and encouraging questions and discussion.
- Be honest when introducing the scheme. Admit to the potential problems and costs and ask for suggestions to overcome them.
- Aim for cooperation not submission (i.e. inter-dependence not dependence).
- If people are raising objections or making suggestions, they may have a good point. Be prepared to listen to them and, if possible, adopt their ideas.
- Remember that change takes time. Don't expect everything to happen at once.
- For change to be effective it should be introduced and implemented in a positive atmosphere of cooperation.

There are two further theories we would like to offer, which expand on the points listed above and give insight into maintaining effective and positive communications. We find they are useful to the change agent and to the participants in the change programme. Both are included as a standard part of our observer training.

Gaining cooperation or getting to 'win–win'

For the behavioural approach to work effectively it needs to be introduced within a cooperative environment. Unfortunately, human beings often become very competitive, particularly in situations where they feel threatened. Competition will lead to low trust and, by definition, low cooperation. For example there is often rivalry and low trust between shifts and between shop-floor and management. In most (although not necessarily all) such situations this leads to lowered effectiveness. It will certainly lead to a less than smooth introduction of the behavioural approach. This will be made worse if there is suspicion concerning the activities of the consultants involved. It is important, therefore, to be aware of any signs of competition and to know how to deal with it, or better still, to know how to avoid it in the first place. There are number of factors which determine how well people work together. These apply whatever the context. The following three are of particular importance:

1 How much people *trust* each other: The level of trust can vary from very high to very low.
2 How *open* people or groups are with each other: Again, openness can be high or low and involves:
3 Whether there is *cooperation* or *competition*. There can be high cooperation or high competition.

All three of these factors can have significance for the relationship between the participants and the consultants or whoever is implementing the scheme. They are also important aspects of the relationship between managers and subordinates in the organization. Depending on past experience of consultants, there may be low trust in outside experts. We mention this as an issue in the case discussed in Chapter 4. It is also not uncommon for levels of trust between managers and shop-floor to be low. The issue is not only how much people believe what is said, but also how much confidence they have that others will deliver on what they have promised. For a variety of reasons managers may not communicate openly and so subordinates feel that they are not getting the full story. Often they have evidence for this. Competition between different departments or shifts is also a common phenomenon in organizations. For any change intervention to be effective it is obvious that all three factors should be as high as possible, between all concerned.

A further complication is that the three factors interact with each other. If trust is low, people tend to be less open and do not communicate and so become suspicious of each other's motives, this leads to low cooperation or competition. This, in turn makes the participants more suspicious of each other and the interaction goes into a downward spiral (or vicious circle) of lowered trust and communication and increasing competition. This is a not uncommon experience in times of change in organizations. If trust is high, people tend to be more open, this openness creates more interaction leading to greater trust, which in turn increases cooperation. This raises the level of trust still higher, leading to an upward spiral (or virtuous circle). It is important to be aware of these processes during a change intervention and ensure that a virtuous circle is developed. The best advice on this is to maintain complete openness as this is the only way to build trust and reduce competition.

In case it should be assumed that we are advocating a 'love and trust' model as a panacea for all ills, it is worth pointing out that we are not saying that openness and trust are appropriate in all situations and that one should always go for cooperation. Whether cooperation or competition is most appropriate depends on the type of situation – whether it is *zero sum* or *non-zero sum*:

- A *zero sum* situation is one where one person's losses are the other person's winnings and vice versa.
- A *non-zero sum* situation is one where there is some outside influence which can add to, or subtract from, the total gains available.

In zero sum situations competition can be appropriate. In non-zero sum situations competition will always lead to both sides losing (*lose–lose*). Cooperation will lead to both sides winning (*win–win*). Most 'real life' situations are non-zero sum. If people cooperate, both sides gain. This applies to everything from war to trade union negotiations. In almost every industrial dispute that has reached the stage of strike action, both sides have lost more than was gained by the strike. In the final settlement, the strikers rarely recoup their losses in terms of lost pay, or if they do it takes a considerable time. Management also loses heavily in lost production and subsequent lost profits. If they could have cooperated and reached a quick agreement both sides could have gained. Similar considerations apply to change initiatives. If a spirit of cooperation can be maintained a successful conclusion can be reached quickly and with minimum cost. If low trust and competition develops, time is wasted and the whole project can be jeopardized.

Most real-life zero sum situations are relatively trivial. Gambling games between friends provide an example. (Gambling at casinos is usually non-zero sum, since there is often a slight built-in advantage towards the casino.) In the situation between friends competition is appropriate. There is, in fact, little point in doing otherwise. However, our advice is not to get involved unless you are better at the game than the other person; otherwise you will surely lose. We know of no such situations within the work of organizations.

Staying OK

Whether the individuals involved in any project start off, and continue with, a positive, open and trusting relationship will depend to some extent on how they feel about each other. There is a framework which we have found helpful in this context. It is derived from the work of Franklin Ernst (1971) and is concerned with how one perceives other people. It is also involves attitude to others, but in this case, since we are referring to our own attitudes, it is possible to do something about them. It is part of a larger theory, but in this context we refer to the concept as 'Staying OK'. In this framework the term 'OK' is used in a semi-technical way, but the meaning is much the same as when we use the expression in everyday life. If someone is seen as 'OK', then they have value as a person and their views and opinions have to be considered seriously. They are competent and are in control of their life. Someone who is seen as not 'OK' is, on the other hand, seen as the opposite of this. They do not seem to be in control, and their views and opinions can be dismissed without consideration. As a person they are of no consequence.

The concept of being 'OK' or 'Not OK' can be applied both to oneself and to others. Thus I can feel 'OK' or 'Not OK' and, likewise, can see you as 'OK' or 'Not OK'. This gives us the four positions shown in Figure 6.2. We will consider each of the four positions in turn. It is important to remember that these positions are *assumptions* that we are making about ourselves and about others. They do not, necessarily, reflect reality.

I'm OK You're Not OK	I'm OK You're OK
I'm Not OK You're Not OK	I'm Not OK You're OK

Figure 6.2 *Staying OK*

- I'm OK, You're Not OK: If I'm OK, then I have value as a person and my ideas and views also have value. You, on the other hand, being not OK, have little or no value. Hence your views don't count and I can ignore them. In extreme cases I can get rid of you by, for example, firing you (or getting you fired). I will feel important and self-righteous, you will feel belittled and discounted.

- I'm Not OK, You're OK: This is, in many ways, the reverse of the above position. Everyone else, but me, seems to be in control and know what they are doing. Whereas I am completely confused and do not know what is going on. I feel worthless, while everyone else seems competent and valuable. In this situation, most people keep quiet and make no contribution, hoping that no one will notice them. In extreme cases they will leave the situation, either temporarily by walking out, or more permanently by resigning.
- I'm Not OK, You're Not OK: In this situation I do not know what is going on and on one else seems to either. We are all confused and since no one understands what is going on there is no way out of the mess we are in. I feel worthless and no one else is competent or of any value. Most people feel this way, at times, in most organizations, particularly in times of change.
- I'm OK, You're OK: In this position I see myself as competent and in control, and see you as the same. We may have different views and beliefs, but I see yours to be equally as valid as my own. If we discuss and share ideas we may come to an agreement, or feel comfortably able to differ. Either way, we will respect and value each other's views.

Hopefully, most people spend most of their time in 'I'm OK, You're OK', but most of us can recognize that there are times when we move off into one of the other positions. The commonest fallback position is, probably, 'I'm OK, You're Not OK'. This is certainly very common in organizations. The implication of this is that when things go wrong it is the other person's fault. The effect of communicating from this position is usually to push the other person into the same 'I'm OK, You're Not OK' position, so that you end up in an argument. It is worth considering what are your own favourite fall-back positions, whether you do move away from 'I'm OK, You're OK' and what are the implications of communicating from such a position. The important thing to remember is that the only effective position, from which to communicate is 'I'm OK, You're OK'.

Unfortunately, it is not uncommon for managers to work from 'I'm OK, You're Not OK', in relation to subordinates (or colleagues). Even more unfortunately, the same can be true of trainers and consultants. It is obvious that this is not a good position to work from, as the other feels put down and belittled. Such an approach will inevitably generate antagonism and start the downward vicious spiral outlined above. It is, however, all too easy to do. Both the 'Not OK' positions in the lower half of Figure 6.2 are also, fairly obviously, not good places from which to work. Clearly, if openness, trust and cooperation are to be generated and maintained, it is essential to maintain a relationship characterized by 'I'm OK, You're OK'.

To summarize the message of this chapter, the reason, we believe, that this approach to continuous safety improvement is so successful is because we start from the position of recognizing and accepting the dependence which frequently exists in organizations. By a process of training and persuasion using (initially) volunteers, we gradually get quite junior members of the organization taking responsibility for measuring the level of safe behaviour and encouraging improvement. We are thus working from dependence towards inter-dependence, via mutual dependence. This process is facilitated by the structure of the programme, which provides a safe environment for participants to experiment with the process of taking control. Being based on well-tried and scientifically established principles (behaviour modification), and introduced in accordance with well-under-stood theories of change, leads to success, which then breeds further confidence and the whole experience becomes self-reinforcing.

In the next, and final, chapter we look at some applications of the behavioural approach in areas other than safety in organizations. These cover safety in wider contexts and also other applications, such as quality improvement within organizations. They are drawn, both from our own experience and from the work of others.

7

BEYOND SAFETY: WIDER APPLICATIONS OF THE BEHAVIOURAL APPROACH

So far our focus has been on the application of OBMod theoretical principles to the improvement of safety in the workplace. However, we have subsequently adapted this model and applied it to quality improvement. It is, in fact, our contention that this same model can be used in any setting where it is possible to differentiate behaviour in terms of some presently exhibited and undesirable behaviour (for example, an unsafe or incorrect practice which is not wanted) and a desirable or required way of behaving. (We give examples of some of these later in this chapter.) The need continuously to improve 'quality performance' would seem to be as much about behaviour on the part of the individual, team or work group as 'safety behaviour'. Indeed, we have found that quality improvement was one of the natural 'spin-off' benefits gained from the behaviour-based approach to safety, due mainly to the raised levels in housekeeping and hygiene standards in the work environment. Since other authors, for example Krause and Finley (1993), have described the concepts of 'safety' and 'quality' as 'two sides of the same coin', this also affirmed our beliefs that the safety improvement model would work to improve quality in much the same ways.

A Behaviour-based Approach to Continuous Quality Improvement

Early laboratory and field study work carried out by Adam and Scott (1971), Adam (1975), McCarthy (1978) and Kim and Hammer (1976) had already demonstrated the potential of behaviour modification techniques in a quality improvement process. These studies, however, tended to focus on specific, time-bound experiments within a particular setting or occupation. We were much more interested in applying a behaviour-based approach within a plant operating a continuous production process, involving several departments, and consisting of multiple skills and skill levels. So, it seemed to us that 'quality behaviour', unlike 'safety behaviour', would be contingent on interactions with other people and we would need to optimize conditions within the working environment to facilitate this, in order to ensure a successful outcome. In addition, it was essential that any new system or process was durable and sustainable by the workforce with minimal input from external consultants. This meant that certain modifications to the basic programme would be needed. In the following section of this chapter we describe the implementation of a quality improvement process – with roots in the principles of behaviour modification – in a production environment, and the modifications that were made to adapt the programme to the needs of an integrated quality improvement process. The initiative took place in the same cellophane production plant we described in the case study in Chapter 4. The basic theoretical principles and practical applications remained the same as in the safety programme, in that it was still a 'shop-floor' driven process which incorporated the concepts of performance measurement, goal-setting and performance feedback. However, it also included a problem solving paradigm aimed at removing or reducing barriers to improving quality performance. We also had to ensure that the structure and climate of the organization were conducive to the changes necessary to facilitate this approach to quality improvement.

We found that the following principles were key to the successful implementation of an OBMod approach to quality improvement in a continuous process production environment:

- Open acknowledgement that an OBMod approach to quality improvement must be regarded as a process not a project.
- Delivery of a clear and consistent message about quality.
- Implementation of a quality opinion survey.
- Use of an upward-driven quality improvement process.
- Commitment to the quality improvement initiative.
- Employee participation through the use of an internal customer model for quality improvement.
- Quality behaviour facilitated through quality improvement teams and the use of 'link-pin' structure.

Each of these is discussed in turn after we examine the need to continuously improve quality performance in an organizational environment.

The need to continuously improve quality performance

Increasing competition from both home and overseas markets and the desire to improve in order to meet customer demand in high volume and repetitive manufacturing has prompted a strong surge of interest in the concept of 'quality through process improvement'. Evidence suggests that a focus on quality and its continuous improvement has resulted in desirable benefits and British industry has begun to recognize that 'quality improvement makes money', rather than the more traditionally held belief that 'quality costs money'. Quality improvement has many advantages

(Feigenbaum, 1986; Hayes and Clark, 1985; Schonberger, 1982; Wheelwright, 1981), including:

- higher profit margins
- lower manufacturing costs
- improving product appeal
- on-time deliveries
- increased direct output
- larger share of the market
- increased ability to attract high-quality customers
- facilitating, building and maintaining a competitive position
- reduced waste/scrap
- reduced cost of repairs
- expanded research and development activities
- improved quality of work life for staff.

Lost business, down-time, rework, the high costs of waste and scrap, re-test and excessive inspections are some of the tangible penalties of poor quality which can lead to the ultimate business failure and demise of a company. In a climate characterized as a buyer's market the customer is king. Customers are looking for value and this means quality and reliability at a reasonable and competitive price. It has been suggested that quality improvement is only achieved when we develop a constructive level of dissatisfaction in an organization and when the total workforce believes that they can, both, be and do better. The challenge is to change individual behaviour so that quality is the paramount consideration in each decision, at all times, at all levels. Nevertheless, it has been suggested that 'quality' is not an issue that is put first and foremost by the workforce because it is not 'rewarded' by management (or the business world). Frank Squires (1980), a quality consultant, said, 'Management is not against quality. Quantity just has a higher priority.' He suggested (and we would agree) that the questions most asked by management are, in order of frequency:

- How many? (quantity)
- How much? (cost)
- How good? (quality)

Employees thus recognize the standards that become 'acceptable' in the company and so perform accordingly. In simple terms, they are continuing to do what they get rewarded for doing. Even if it appears that some form of punishment is likely for poor quality behaviour, it is unlikely to have much, or any, impact on the behaviour of the individual, because he or she knows they are not likely to get caught, and/or punished, every time they produce a poor quality product. Thus, the cutting of corners and the concealment of errors and mistakes will continue and result in negative impact on quality performance. If error-producing behaviours, poor workmanship or unsafe practices are accepted (i.e. condoned), they thereby become reinforced and evolve as the accepted norm within the company. Changing to a more positive organizational culture takes time and is difficult to achieve. Thus, some quick-fix project to improve an aspect of poor quality might have an immediate and short-term impact on the problem but is most unlikely to make any difference to the organizational climate and culture. To do this, an approach is needed which will become integrated as part of the process within the organization.

Open acknowledgement that an OBMod approach to quality improvement must be regarded as a process not a project

Continuous improvement requires a constancy of purpose, the development of long-term strategies and an overtly expressed high level of commitment to the purpose and strategies formulated. Therefore, the behaviour-based approach to quality improvement should be introduced as part of the processes and systems used within a plant or organization and regarded as 'the way we do things here', rather than as one of the 'start–stop' projects or programmes which seem to abound in organizations, and which are the cause of 'change fatigue' commonly observed in contemporary organizations. Employees will often perceive such initiatives to be simply paying lip service to an edict from some remote parent company or senior management; or perhaps that the company is signing-up to the yet another 'flavour of the month' cure for organizational problems because everyone else is doing it! This leads to resistance to the change and the initiative will suffer from a lack of credibility and ownership. Employees may react by attempting to ignore what is going on, on the grounds that if they ignore it long enough it will go away. Sadly, the change makers know that they are likely to be correct in this wisdom. Research experience suggests that it is relatively easy to initiate a quality improvement programme and show some noticeable and significant improvement in quality performance. However, these are usually short-term results and conditions quickly return to the status quo as soon as management attention is focused elsewhere. Sustaining such a performance is much more difficult and it requires both employee ownership of the process and their acceptance of the need for change if it is to be durable and successful.

Our target organization acknowledged these concerns. They believed that they were good at starting new initiatives but found it difficult to sustain them over a long period of time. Other quality improvement programmes had been introduced by management with some success, but these had been short-lived, to become memories that were likely to taint employees' impressions of any new approach to quality improvement. So, to a certain extent we were ready for such comments as, 'We have done that before . . . and it didn't work . . . it's a waste of time'. The management of the company also made us aware of the strong need to move from a focus on a high volume of product towards the targeting of specialized, high quality, high cost products. Although this message had been delivered to the workforce many times, it was feared that behaviour still had not changed towards meeting this new business objective.

One very effective way to raise quality consciousness and determine the needs of a quality improvement programme is through an opinion survey. Thus, this was used as a start in the process of gaining ownership and commitment to the changes that were to be made. Other ways of helping the workforce to gain ownership of the process were built into the fundamental design and these will be described later in the chapter. Nevertheless, before the opinion survey could be carried out a clear letter of intent had to be delivered to the whole workforce.

Deliver a clear, consistent message about quality

Kaeter (1992) suggests that clear, consistent communications are vital. This includes communicating the expectations that management has of each employee; giving honest answers to concerns; maintaining a continual dialogue between all levels of management; reassuring employees that they can do the job; and facilitating the transfer of responsibility to workers by coaching and giving support. This employee ego development is the basis for empowerment which is fundamental to an upward-driven organizational change initiative. Management commitment to quality must be seen to be of paramount importance at all times, by words and by deeds! It is essential that this message is endorsed at the highest possible level. Therefore, the first step in the OBMod approach to quality improvement is the delivery of a letter of intent, from the chief executive or most senior management figure, to all employees. Typically, this letter should include:

- A statement about the current organizational situation and economic climate.
- Why 'quality' is so important to the company. This should be a personal perspective, which describes why quality must continuously improve – for example, the costs of waste or late orders; the strategy needed in order to gain new customers; or perhaps why and how the loss of a key account, due to poor quality product, would be so disastrous, must be explained.
- What is in it for the employee.
- What can be done (the long-term quality plan) – a brief description of how the OBMod approach works in practice.
- Notification of those persons who will be the champion of the process, or the persons responsible for the implementation of the approach; and/or an introduction to the external consultants being used.
- Who will be involved. Whilst quality is the responsibility of everyone, it is likely that an organization will find it most beneficial to begin the process in one part of the company and then gradually include other departments or areas. So, people need to be told if the initiative is going

to start as a pilot scheme, or if it is going to begin in one department or area and then slowly spread to others. This should avoid the problems of people feeling that they are either being singled out or left out!

- How it will work – a brief step-by-step account of the implementation process, which explains the survey process, the contributions to be made by the interviewees and an advance notification that the opinion survey will be the starting point for most employees.
- When it will happen, i.e. a timetable of events.
- It is also useful if the letter is used to anticipate and answer some of the questions that are likely to be asked. For example:

1 Why are we using this approach rather than something else?
2 Will it cost a lot of money?
3 Will it affect jobs? (If possible, give some reasonable assurances about job security.)
4 Will it mean more (harder) work for me and will I get more pay?
5 How quickly do we see a result?
6 Will I see all the results of the survey or just the ones the management chooses to show me; will it really be anonymous?

The letter must be a statement of policy which clearly and concisely defines what is expected of all employees with regard to quality performance standards. It is an endorsement of priority, meaning and evidence of commitment.

Implementation of a quality opinion survey

The key to quality improvement is measuring the current situation and then establishing a process that will effectively raise that level by the elimination or minimization of the identified problems. This means that quantified information must be available to evaluate the results of any quality improvement activity. The importance of measurement cannot be overstated:

> If you can't understand something you cannot measure it. If you can't measure it you cannot control it. If you can't control it you cannot improve it. (adapted from Harrington, 1987)

For these reasons an important part of any quality improvement project is an anonymous quality opinion survey. The aims are to:

- Specify where the company is (i.e. the current position).
- Identify the actions that are required to build and sustain a quality improvement process.
- Identify opportunities.
- Anticipate problems before they occur.
- Listen to ideas and suggestions for quality improvement put forward by all employees.

Feedback of the results of the opinion survey, to all personnel, provides an opportunity to sow the seed of commitment to a behaviour-based approach to quality improvement. It also signals management commitment to the importance placed on quality and the need for quality improvement, and provides a forum in which to exchange suggestions and ideas. In essence, it can facilitate the process of employee involvement and empowerment (McAfee and Winn, 1989).

Both interview and questionnaire techniques were used in this survey. Fifty, semi-structured interviews were conducted on-site with a randomly selected, representative, stratified sample of employees. (See Appendix B for an example of some of the prompt questions used in the interview; note that this can only be developed after some open interviews/discussion sessions are completed.) A questionnaire was generated from the interview data and distributed to 520 staff to be completed anonymously. This produced a 45 per cent usable response rate for analysis. The objectives were to gain an understanding of the perceptions of quality within the company and readiness for change; to provide baseline data for evaluation purposes; to identify quality improvement opportunities and potential barriers to change; and to help management develop a sense of awareness about quality and quality improvement needs in the plant. The questionnaire included measures of job satisfaction (Warr et al., 1979), organizational commitment (Porter et al., 1974), perceptions of cooperation and morale in the plant and measures of certain job characteristics associated with quality work performance, such as skill variety, autonomy and feedback (Hackman and Oldham, 1975).

The results of the survey indicated a high degree of acceptance (93 per cent) of the importance of external quality in relation to the future of the company, but complacency about external quality was seen as a potential barrier to quality improvement (that is, 'external quality is good, therefore there is no problem'). A limited knowledge existed about the actual quality of the product, external customers (i.e. who are our customers?), the costs of waste and returns and the competition. Only half of the respondents believed that internal quality was good. It was recognized that ways of improving inter-departmental knowledge should be explored (for example, talking to each other at machine operator level, job exchange, visits and shadowing). Indeed, the company's policy on quality was poorly understood and it was believed that the quality message was still regarded by many employees as mere lip service. Lack of cooperation in the plant was regarded as a barrier to doing the job properly and to quality improvement. The need for more timely feedback was reported as a factor in quality improvement by over 80 per cent of the survey respondents. Still, it was encouraging to note that the suggestions for improving quality in the plant were not just restricted to the need for more investment in plant and machinery, replacement and repairs, but included aspects of employee behaviour. There existed a real desire to be more involved and aware about customers, and to find out what happened in other parts of the plant.

The findings supported our plans to improve inter-departmental coordination and cooperation by involving all personnel in identifying their 'internal customer' requirements, their own needs from suppliers (internally and externally), and the quality behaviour associated with these. The results helped significantly in the development of the behaviour-based approach to quality improvement and in defining the next steps to be taken. Verbal feedback of the results of the survey was given to small groups of employees. It was necessary to ensure that there was enough time to discuss these results and listen to the comments and ideas from the workforce. As promised at the outset, access to the full written report was given to any interested employees.

Use an upward-driven quality improvement process

Typically, quality improvement in contemporary organizations has been addressed through the introduction of concepts such as 'quality circles', team building/team development events and 'empowerment initiatives'. There is a plethora of books that offer guidance and often the strategy adopted requires the purchase of an 'off-the-shelf' package. Varying degrees of success with these are reported and the reasons for this are as varied as the systems or packages themselves. Most seem to focus on the management structure of the company and are top-down directed. We fully accept that management must lead an initiative, but quality is everyone's responsibility, and so following the successful application of the behaviour-based approach to safety using an upward-driven improvement process, we believed that this same approach could be modified to fit the quality improvement process.

In the accident prevention process efforts are concentrated on the modification of behaviour to avoid injury; similarly, in quality improvement it is necessary to direct behaviour towards the avoidance of errors and defects – that is, the principle of right first time. It means that problems must be resolved 'upstream' rather than the more costly process of detection of quality problems 'downstream'. Early studies at Hewlett Packard indicated that a defective resistor cost 2 cents if it was thrown away before use, 10 dollars if found at board assembly level, and hundreds of dollars if it was not detected until it reached the customer! Since those closest to the work may be best able to make suggestions about quality improvements and identify the behaviours associated with good versus poor quality performance, the emphasis must be on employee participation, mutual problem solving and team working. To the extent that product defects are the result of poor workmanship, quality performance can be influenced by behaviour modification techniques. Furthermore, it requires that employees liaise effectively with their internal suppliers and customers to establish what such behaviours might be. Dale and Cooper (1992) suggest that 'Quality should simply be the way of everyday, organizational functioning and is *the* way to do business', and so the success of a behaviour-based approach lies in employee participation, mutual problem solving and team working. Therefore, the focus is on the employee (and/or team), who is determining quality before and during operations, rather than after the materials have been produced.

Quality consists of two elements: quality of conformance (have we met specifications?) and quality of design (are the specifications correct?). In the case of the behaviour-based approach, such specifications are expressed in terms of specific, observable and measurable behaviours. Rather than being a somewhat abstract truism, quality is, in effect, defined for sub-units of the overall process in terms of specific units of behaviour. The previous use of this approach for safety improvement suggests that it would work equally well for quality of conformance. Quality of design, however, is dependent on increasing levels of inter-departmental communication and the establishment of an internal quality model. Therefore, the approach should facilitate both a team model and an internal customer approach to quality improvement. Before this structure is discussed in detail, it is important to consider the issue of gaining commitment to a quality improvement process.

Gain commitment to the quality improvement initiative

Many organizations have described the improvement of quality perform-ance as one of their corporate plan key objectives and they typically produce a mission statement which describes this as a goal and aim for the business. Unfortunately, much of this simply does not translate into the day-to-day activities of employees, particularly for many of those working at an operational level known as 'the sharp-end'. We believe that this mission statement, written in global or generalized terms, rarely touches the first line of thought amongst managers or the workforce as they go about their daily business. These well-intentioned, embossed certificates displayed on notice boards are rarely 'seen' in any real sense and thus become 'wallpaper'. Maybe they are not read; maybe they *are* read but are not acknowledged, absorbed or believed to be relevant to the reader. Most probably, however, they are not recognized or understood as a communication or instruction which can have any direct impact on the immediate, day-to-day behaviour of the individual employee. If you are feeling brave or confident that these assumptions are wrong, try asking some of your employees, at random and without warning, what the mission statement says and what it means to them as an employee of the organization! Whilst we accept that it is unfair to suggest that quality improvement in organizations stops at the mission statement and acknowledge that considerable effort is expended on quality improvement initiatives, we also contend that lack of ownership for a programme of change is likely to stem from poor communication of the need for change and lack of information on what employees are expected to do differently, in order to effect change towards an improvement in quality performance.

To overcome this potential problem, part of the training programme included a session on producing a personal mission statement for each group (for example, by department or shift). The company policy statement on quality was reviewed and used as a basis for the development of these local, quality policy statements. These had to be written in behavioural terms, which were specific, observable and measurable, and were displayed in the departments concerned. Ultimately, the group members would use their own quality policy statement to answer two questions:

1 In specific, observable and measurable ways how must I behave today (or on this shift) to meet the quality specifications agreed between me and my customers?
2 In specific, observable and measurable ways how must I behave today (or on this shift) to ensure that I can be assured of the quality specifications agreed between me and my internal suppliers?

The concepts of employee participation and the use of an internal customer model to achieve quality improvement are discussed next.

Employee participation through the use of an internal customer model for quality improvement

Even in continuous production process companies the typical organizational chart shows a series of separate, vertical chimneys (Nichols, cited Kaeter, 1992). In reality the work moves horizontally through the organization from one function to the next and most quality problems exist at cross-functional points that the product must pass through en-route to the external customer. Nichols advises that we should teach people within the organization to treat each other as internal customers. In terms of the behaviour-based approach to quality improvement it means that the internal customer requirements are interpreted in the form of specific, measurable and observable behaviours which are reinforced through the medium of informational feedback. Each function within the plant has its own customer(s) and supplier(s). Focus discussion groups (i.e. departmental quality improvement teams) are used to identify desired behaviours and the barriers and inhibitors to quality behaviour in each of the departments. However, one group of individuals should oversee the total process by acting as the quality improvement steering group.

The purpose of the steering group is to oversee the development, implementation and monitoring of the quality improvement process across the total workforce; to identify best practice in the quality improvement process; to guide and support the departmental quality improvement teams; to sustain the process as a long-term strategy within the company; and to adjust the process as required to meet the changing needs of the organization. Membership of the group consists of representatives from all key departments and functions, with appropriate management, union and employee representation. Since this group could be large, that is, approximately 18–20 people, it is suggested that a smaller core group is identified, which meets often and regularly, and where other members of the group attend on a need basis. A key member within this group is the 'quality improvement facilitator or champion'. The formation of this group structure utilizes a link-pin, as shown in Figure 7.1 (Likert, 1961). This assures continuity in communication and decision making across groups by means of individuals who are members of more than one group – for example, the group that they work in (e.g. the department) and a higher-level group of which they are also a member and a representative of their own departmental group. Communication occurs laterally and vertically, and knowledge and decisions are transmitted by the means of overlapping group membership. This tends to result in high levels of productivity, performance and member satisfaction.

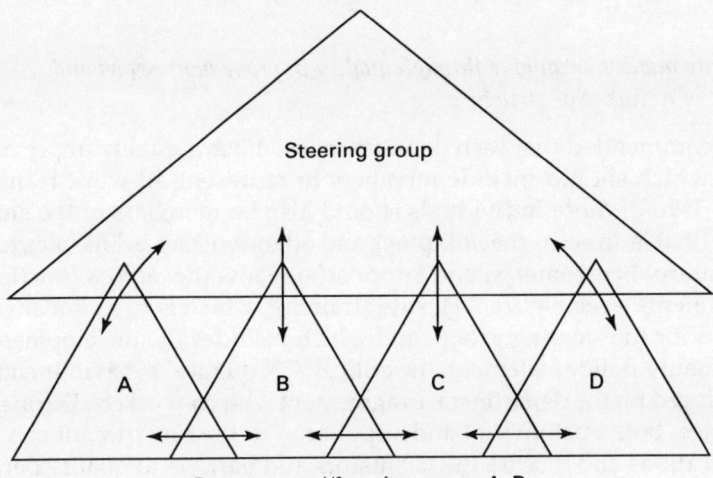

Notes: The arrows indicate the link-pin function
Teams B and C display a full supplier and customer link-pin function

Figure 7.1 *The link-pin function*

Training for the steering group should incorporate the following elelments:

- An understanding of quality improvement issues and how these relate to interactions between departments in the plant (the internal customer model).
- An understanding of the behaviour-based approach to quality improvement and the strategies used.
- Training in relevant process skills (e.g. team building, working in groups, decision making and the management of change).

- Interpersonal skills training to facilitate communication between individuals and departments.
- Sessions to anticipate and solve potential problems in the application of the behaviour-based approach, and to identify the facilitators and barriers to quality improvement.
- A session to produce the steering group's quality policy statement in order to encourage ownership of the process.
- The development of checklists, that is the identification of critical quality-related behaviours.
- Training in the mechanics of the use of the checklists, and in the methods (and importance) of the provision of both visual and verbal feedback.
- Training in the setting of goals/targets.

After training, members of the steering group will be ready to facilitate the development of the departmental quality improvement teams.

Facilitate quality behaviour through quality improvement teams and the use of a 'link-pin' structure

It is recommended that each department establish a quality improvement team, which should include members to represent all work teams and shifts. Two of these individuals should also be members of the steering group (that is, to act as the link-pins), and others will act as link-pins to their own internal customer(s) and supplier(s) teams (i.e. across functions or departments – see Figure 7.1). After training, which is to a similar design as used for the steering group, and which includes the development of a *team* quality policy statement, the critical 'OK quality' behaviour checklist is produced for the department, in agreement with co-workers. During team meetings, both customer(s) and supplier(s) to the department can agree specifications and discuss the facilitators and barriers to quality perform-ance. Regular meetings of the departmental teams are needed to discuss quality behaviour and the reasons for 'NOT OK quality' behaviour. That is, the groups should have an opportunity to identify the potential barriers to quality performance. Problems and issues that cannot be resolved at this departmental level are passed upwards and dealt with by the quality improvement steering group. By these means it is possible to address the criticism that a behaviour-based approach cannot be applied to the macro issues characteristic of complex organizational change efforts.

Within our target plant, a pilot project was introduced into a single department with the intention of gradually developing and spreading the process plant-wide. As with the behavioural approach to safety, this is a practice we would highly recommend since it is possible to learn from mistakes before they become too costly or difficult to put right. Our pilot group produced a checklist and the observers obtained a four-week baseline measure of 51 per cent 'OK quality'. A target of 70 per cent was set for quality improvement and over a 12-week period the average quality performance was 79.8 per cent and peaked at 92 per cent at weeks 11 and 12. Many issues were raised and resolved by the group members at the regular departmental team meetings. This was facilitated by the presence of a 'link-pin' representative of 'supplier' or 'customer' departments. Other ideas were passed to the steering group for further discussion and decision. The motivation to improve quality remained high and commitment to the approach was evident from the efforts made and the improvements measured. The changeover of observers for the next phase renewed this enthusiasm and brought in new blood with fresh ideas, thereby ensuring continuous quality improvement.

An overview of the OBMod continuous quality improvement model

Essentially, a five-step process of behaviour change, as described by Luthans and Kreitner (1975, 1985) was used as a basic model for this intervention. This includes:

- Step 1 Identify the key behaviour (i.e. produce a checklist for quality behaviour).
- Step 2 Measure the baseline frequency of the key behaviour.
- Step 3 Functionally analyse the behaviour to identify the antecedents and consequences.
- Step 4 Intervene by changing the antecedents and consequences to increase the frequencies of the desired behaviours and decrease the frequencies of undesirable behaviours.
- Step 5 Evaluate to determine whether or not the behaviour and performance have changed in the desired direction.

However, in order to develop this process for use in improving quality, more steps were incorporated into the basic model. Table 7.1 gives an outline of all the stages involved.

Evaluation of the behavioural approach to quality improvement

It is our experience, from this intervention, that defining 'OK Quality' behaviour in behavioural terms and then positively reinforcing it is a viable approach to quality improvement, but it must be supported by a structure and climate that facilitates the changes necessary to sustain a quality culture, such that quality becomes part of 'the way we do things around here'. There is very little doubt that the fact that we had previously carried out a very successful behavioural safety intervention in this same organization contributed to the success of the quality improvement programme. Participants were already very familiar with, and committed to, the behavioural approach. A further effect is that this methodology actually facilitates genuine empowerment. Those taking part had already had experience of taking responsibility for their own behaviour in relation to safety, so that it was relatively easy to extend this to the wider problems of

Table 7.1 *Steps to continuous quality improvement – an OBMod approach*

1 Letter of intent to employees from the most senior level of management

2 Open and semi-structured interviews with a random sample of personnel

3 Design and pilot a questionnaire for the quality opinion survey

4 Conduct the opinion survey – feedback survey results

5 Produce (if not available) or amend the company, quality policy statement

6 Establish a Quality Process Steering Group

7 Train the Steering Group

8 Establish the Departmental Quality Process Improvement Team (with 'internal customer and supplier' link-pin representation)

9 Train the Departmental Process Improvement Team; identify observers within each team or group (the same process as for the safety improvement model)

10 Produce critical behaviour checklist inventories; these are written in the same format as for the critical safety behaviour checklists but using the terms 'OK Quality Behaviour' and 'NOT OK Quality Behaviour', rather than 'SAFE' and 'UNSAFE'; practise using the checklists and agree final format. Display large-scale copies of the checklists where they can be seen easily and regularly

11 Establish a quality performance baseline measure over a minimum period of four weeks

12 Use the baseline measure to set a target for quality improvement; these should be participative goal-setting sessions rather than a goal which is assigned by the observer

13 Continue the intervention for a period of 12–14 weeks; the cycle for this will depend on the shift pattern within the organization

14 Monitor the intervention, evaluate the results and provide feedback

15 Discuss outcomes and modify the process as necessary; plan the introduction of the approach to additional departments

coordination and control involved in quality. In other words, we had, to a large extent, already overcome the problem of dependency (as discussed in Chapter 6).

We have discussed in some detail this one intervention in which we were personally involved. There are, of course, many other applications of the behavioural approach, both in organizations and at a more personal level. We will outline some examples of these in the rest of this chapter.

Other applications of the OBMod Approach

Absenteeism

Apart from safety and quality, behaviour modification has been applied in many other areas of organizational behaviour. For example, OBMod has been successfully used in dealing with absenteeism and lateness. Often it is assumed that absenteeism is a symptom of dissatisfaction with some aspect, or aspects, of the job. The solution, therefore, is to improve job satisfaction. This may, indeed, have some impact, but there is often a limit to what can be done to improve satisfaction. Shift work still needs to be worked, and many other 'dirty', and undesirable, jobs need to be done. In addition, the rewards for staying away from work are powerful, and rarely capable of being influenced by management. OBMod, on the other hand, looks not at the *influences* that are thought to underlie absenteeism, but rather at the *consequences* to the employee of attendance or non-attendance. It adopts a 'direct action' model.

One approach is to find a means to reinforce the desired behaviour, preferably on a variable ratio schedule. One such example was reported from a factory in Liverpool (see Chapter 3). The introduction of a prize draw, based upon tokens collected for attendance, led to the reduction of absenteeism levels to almost zero. Other schemes have used cash bonuses for those attending on days chosen at random during the month. Both of these types of schemes use variable ratio reinforcement – one by means of a draw, the other by the random choice of the day on which attendance is rewarded. (It is worth noting that both also required attendance at the required starting time, thus also reinforcing punctuality.)

Such examples demonstrate one of the conditions under which reinforcement of this sort works best, that is, the expenditure of a little additional investment on the part of the individual with the potential for a large payoff. Football pools and national lotteries make use of this principle, and it also underlies the use of the prize draws widely used as a marketing technique. Managers sometimes raise (quite legitimate) objections to such schemes on the basis that they involve paying workers extra to do what they are already paid to do (i.e. come to work), but sticking to your principles may have a cost – in this case continued high levels of absenteeism while these well-proven behavioural techniques could solve the problem.

It is perhaps worth noting in passing the contrast between schemes that reward attendance, and those that punish absenteeism. Schemes that use punishment usually do so by giving an attendance bonus and then removing it for an absence. (On initial inspection the attendance bonus may sound like a reward, but as we will see, this is not always the way it is perceived.) For example, managers recently tried to improve the attendance of a group of UK Civil Servants (driving test examiners) by giving a £30 weekly bonus for attendance. The whole sum was forfeited for any non-attendance during the week, no matter what the reason for the absence. The scheme provoked a national one-day strike! One problem with this type of scheme is that once the bonus is lost, during any week, there is no further motivation to attend. It might be worth considering how this, not inconsiderable, sum could have been used to improve attendance by variable ratio reinforcement.

It is also possible to use punishment as a way of influencing attendance. It should be remembered, however, that to be effective punishment should occur after every instance of the behaviour and as soon as possible after its occurrence. These requirements can be compared with the way that many disciplinary schemes operate. Often no immediate consequences follow a period of absence. A record is kept (often by the HR department) but no action is taken until a pre-determined level of absence (often over a period of six months or a year) 'triggers' action by the organization. This then often follows the legally required sequence of verbal and written warnings. The impact of these warnings is, in our experience, temporary. Attendance may well improve over the short-term, but then slips back.

One method of discipline that has been shown to be effective is the use of return-to-work interviews. Immediately the employee returns to work following a period of absence (no matter how short), they are interviewed by their line manager in order to identify the cause/s of the absence. If the cause/s are considered legitimate sympathy and offers of assistance may be in order. If, on the other hand, the causes are not considered valid then informal disciplinary action can be taken. Fowler (1998) suggests that 'return-to-work interviews are probably the most influential element in ensuring absences are not treated casually'. He points out, however, that 'no absence reduction programme can be effective without the full involvement of the line managers'. This involvement is often difficult to achieve. There is a lot of evidence to show that many managers feel uncomfortable carrying out such interviews and will, therefore, find ways to avoid them. It will be necessary to find ways in which the carrying out of these interviews is either less threatening, or even rewarding. Having two managers at an interview may reduce the unpleasantness, although it may overawe the employee if not carefully done. In addition, senior managers should monitor and socially reinforce the carrying out of these interviews. Although such methods may improve attendance, as Landau (1993) has shown, the addition of incentives to a discipline regime will improve its effectiveness. This is consistent with behavioural theory, which suggests that the combination of rewarding the desirable whilst punishing the undesirable can be a very effective way of changing behaviour.

Sales and service

Two studies reported by Komaki et al. (1977), although carried out some time ago, still provide useful examples of how, by using a behavioural approach, the performance of staff in shops can be improved. The first example concerns the performance of two members of staff working in a neighbourhood grocery store. Both the owner of the store and the staff concerned agreed that performance could be improved and agreed to examine how this might be achieved. Analysis showed that when the owner thought that the staff were not performing to the level he expected he would nag them. It was interesting, however, that when the owner was asked if he had ever specifically described what he wanted them to do, he replied that they ought to know since that was what they were being paid to do! The first step, as always, involved identifying the precise behaviour that was to be targeted. As suggested by the theory, this should involve the encouragement of desirable behaviour, rather than the attempted punishing of undesirable behaviour. After some analysis, the behaviours that were identified were:

- At least one of the workers to be within 3 feet of a display shelf, meat counter, or refrigerator when the store was open.
- Any customer to be approached and offered assistance – this assistance to begin within 5 seconds of the request.
- Shelves to be filled to at least 50 per cent of their capacity.

Before the intervention started the level of these behaviours was measured. The level for each of the behaviours was 53 per cent, 35 per cent and 57 per cent respectively. Having identified the behaviours, it was now necessary to specify the consequences. These were: time off, visual feedback, through a graph displayed in the storeroom, and self-recording. The time off was a half-day a week whenever they attained at least 90 per cent on the desired behaviours. The results showed dramatic improvements for all three behaviours. The averages for each increased to 86 per cent, 87 per cent and 86 per cent, respectively. The staff themselves seemed pleased with their improved performance and, although the views of customers were not actively sought, the researchers reported that customers were heard to comment on the improved service.

The second example from Komaki et al. (1977) concerns an employee responsible for running a small amusement arcade who was about to be sacked. Komaki and her co-researchers persuaded the owner of the arcade to give them a chance to try to help the employee improve his performance to an acceptable level. The behaviours that were targeted in this case were:

- Being within 10 feet of the change counter when a customer was within 10 feet of the counter, or when there were five or more people in the arcade.
- Repairing a machine when it malfunctioned, or refunding the customer's money, as appropriate.
- Cleaning the premises or the machines when fewer than five people were in the arcade. Five areas were identified which had all to be cleaned over the five days of the working week.

At first it proved difficult to find appropriate reinforcers. The owner refused to consider extra pay, time off, or even social reinforcement. For his part, the employee refused to consider self-reporting as he considered it to be 'stupid'. Finally it was decided that basic pay would be made contingent on performance of the desired behaviours. At the end of the working week, the hours that attendant had worked was multiplied by the percentage of target behaviours achieved in order to calculate his take-home pay. This placed an emphasis on 'work time' rather than 'clock time'. (Although the attendant's reaction to this scheme is not reported, it may be that this was accepted as the only alternative to the sack.) During the period before the scheme was introduced the attendant's average performance was 63 per cent, sometimes dropping as low as 40 per cent. When the scheme was introduced this increased to 93 per cent, and later to 97 per cent.

In reporting the results of these studies the authors were at pains to note that it was not only the behaviour of the employees that changed – so did that of the employers. In both cases they had to define, and communicate to the relevant employee/s, the precise behaviours that were required. In addition, the requirement to monitor the desired behaviours meant that they had to attend to positive working behaviours, rather than ignoring or reacting negatively to instances of undesirable behaviours. Indeed, this is something of which managers should be generally aware – their own behaviour is likely to change as well as that of their subordinates. There is also an empowering effect for the subordinates, because it is now clear what he or she is expected to do.

An example from Hall (1983) shows not only the effectiveness of such schemes in improving sales, but also demonstrates the principle of rewarding the desired behaviour, rather than the final outcomes. The project was designed to get estate agents to sell more. One of the reasons for implementing the scheme was that, although the firm was the largest in the region in terms of staff, it came seventh out of 15 in terms of sales. One possible method of increasing sales is, of course, to reward (reinforce) the achievement of sales targets or, as commonly happens, give a prize to the highest performing salesperson. There are dangers in both these approaches. If sales targets are unrealistic the effect may be de-motivating. If competition is used it may be that it works to the disadvantage of the firm as a whole (for example the withholding of leads that might help other sales staff).

The main target behaviour identified was that of 'face to face' contacts. Such behaviour was considered to be the 'bread and butter' of selling. Initial contacts provide the starting point for a possible sale, and follow-up contacts help cement the relationship. The more of these contacts that could be encouraged, the more likely it was that business would follow. There was also a secondary behaviour that the firm wished to encourage. Attendance at bi-weekly sales meetings was low (approximately 50 per cent), and it was considered desirable to increase the number of agents attending. The method used was to reward the number of initial and follow-up calls made each week. Each agent had to report six initial contacts and 15 follow-up contacts (all 'face to face'). After this target was reached tokens were awarded for additional contacts. These tokens could be exchanged for a range of goods, from a gallon of petrol to a slate pool table. The tokens were distributed at the sales meetings. Although the number of contacts was self-reported by the agents, a check was made through 'courtesy' post-cards and telephone calls. In the year before the programme the firm employed 8 per cent of the agents and had 6 per cent of the market. During the year of the programme the share of the market increased to 13 per cent, whilst the number of agents employed dropped to 5 per cent. In addition, the attendance at sales meetings rose from 50 per cent to 97 per cent. Following the removal of the scheme, however, the sales figures dropped back to below the initial level. As mentioned before, whilst some behaviours become 'self-reinforcing' others may need other techniques to ensure that they are maintained.

Productivity

A nice example of the difference between using fixed ratio and variable ratio reinforcement to increase productivity is provided by the beaver trappers example described in Chapter 3. A more useful example, perhaps, concerns the productivity of cleaners (Hall, 1983). The project was carried out in a large urban hotel in the USA, which employed approximately 100 full- and part-time cleaners, mostly female. Although the rooms were cleaned on a daily basis, it had become apparent to the management that the guest rooms were in need of a thorough, 'deep', cleaning, a task that had not been undertaken for 10 years. At the heart of the scheme was a checklist of the desired behaviours. In total 72 checklist items were identified. Of these 22 covered various aspects of 'complimentary' services such as the number and location of towels, matches and tissues. The rest covered the actual cleaning required and was divided into three categories: 21 items covered the 'bath and tile' areas, 14 covered the wash basin area, and 35 covered the bedroom and living areas.

The deep cleaning part of the exercise was introduced in stages, and was voluntary. Those not wishing to take part were not forced to do so. The first of the three stages was the 'bath and tile' area, followed by the wash basin area, and finally the bedroom and living area. Each of the cleaners nominated a number of target rooms whose 'bath and tile' areas were evaluated, by their supervisor, after cleaning. The areas were scored according to the number of items that were satisfactory. The scores were posted on a bulletin board and credits were awarded according to the scores. A score of 18 or 19 out of 21 received 1 credit, whilst a score of 20 or 21 received ½ credits. These credits could be exchanged for time off, store vouchers, or meals. The same procedure was used for the other two areas and the scheme was maintained by a regular checking procedure. The results of the scheme were beneficial, both for the employer and employees. For the employer, the hotel was deep cleaned for a price that was approximately 25 times less than the cost of a one-off cleaning contract. In addition there was a reduction in staff turnover in what had been a highly volatile workforce. From the cleaners' perspective there were fewer customer complaints and an increase in tips.

It is generally accepted that one of the principal barriers to more efficient production is employee motivation. For this reason many organizations seek ways to increase motivation levels. What is often overlooked, however, is that it is often not the *level* of motivation that is the problem, but *where* motivation is directed. A study carried out by a part-time Masters student, supervised by one of the authors, may help demonstrate this phenomenon. The student was responsible for managing a small section of a utility company. The main duty of this section was to respond to customer emergencies, both large and small. The initial contact point for the customer was a telephone receptionist, whose primary responsibility it was to answer the phone and transfer the call to the appropriate person in the section. It was important that no calls went unanswered so a system was in place such that, if the phone was unanswered for six rings, a loud buzzer sounded. When this occurred it was the responsibility of anyone close to a phone to answer the call. A new receptionist had recently started work and the manager, over a number of weeks, developed the impression that the buzzer was sounding more than it had in the past. In order to check her suspicions she discretely observed what happened each time the telephone rang. To her surprise she found that the buzzer sounded on 49 per cent of all calls. The manager also made a note of what the receptionist was doing at the time the calls were going unanswered. It became apparent that the employee concerned was not 'shirking', but was doing other, but less critical, tasks such as photocopying or filing. Knowing the 'Premack' effect, namely that given a choice people will spend more time on 'nice' rather than 'nasty' tasks (this is a phenomenon first noticed by David Premack – for a good general account see Makin and Hoyle, 1993), the manager realized that answering the telephone must be perceived as a relatively nasty task by the receptionist.

Conversations with the receptionist and other members of the staff quickly revealed the reason for the apparent reluctance to answer the telephone. The receptionist was new and inexperienced and hence often transferred the incoming calls to the wrong member of staff. This often caused annoyance to the members of staff concerned who would often respond with a mild 'put-down'. Answering the telephone, therefore, would often lead to punishment, with no rewards for getting it right. It was decided, therefore, to try to remove the potential punishment in answering the telephone. Extra training was given to the receptionist but, in addition to this, the other members of staff were asked to cooperate in what essentially became on-the-job training. If the receptionist transferred the call to the wrong person the member of staff concerned was asked to transfer the call to the correct person. They were then asked to call the receptionist and, in a helpful manner, explain what they had done and why. As a result of this, the receptionist began answering the telephone much more often.

Personal Applications of the Behavioural Approach

The techniques that we have described throughout this book can also be used to change or improve many aspects of one's own behaviour. As with other applications, it is necessary, first, to specify precisely the behaviour which is required (that is, think about what you want to do, which is different from what you are doing now), and produce an appropriate checklist. It is then quite possible to monitor one's own behaviour, simply by checking at random intervals how well you are performing against the specified criteria. Observations can be done at any time you think about it, or it is possible to set a timing device to indicate the time of the next observation. This approach can be used at home or at work. We have used it successfully in our safety programmes where an individual works on his or her own. We suggest below, with some possible checklist items, contexts where the behavioural approach can be used at home. We strongly recommend trying some of these, but you will need to think and maybe observe yourself for a period to compile appropriate checklist items.

Safety in the home

More accidents take place within the home than in any other location. This is partly because we spend a lot of our time at home, but also because we tend not to think too carefully about safety in this context. There are often many hazards that would immediately attract the attention of a works safety officer. (This is probably true even in the home of safety officers!) Possible checklist items are:

- Stairways and passages to be kept clear at all times (tripping hazard).
- Flexes on electrical appliances to be coiled and kept tidy (tripping hazard).
- All electrical appliances to be correctly fused.
- Saucepan handles to be kept within the area of the top of the stove (danger of scalding).
- Use a step ladder to reach items at a height (do not stand on chairs or other furniture).
- Hold the handrail when using the stairs.

There are also many hazards when working in the garden. Some items to consider are as follows:

- All electrical appliances to be connected via a residual current device (RCD).
- Goggles to be worn when strimmer or power hose being used.
- Stout shoes to be worn, particularly when mowing lawn.
- When not in use, rake left such that prongs cannot be trodden on.
- Ladders to be footed and tied.
- Gloves to be worn when collecting rubbish; pruning shrubs/trees.
- Garden chemicals to be stored out of reach of children/animals.
- Hoses to be coiled when not in use.

The reader could find it a useful exercise to add to and use such checklists in their own work at home.

Improving efficiency or quality in the home

Similar checklists can be compiled to improve performance and quality of output for many household tasks, doing the housework or cooking being obvious examples. Checklist items would include having the correct utensils available, arranged in the most efficient way. This may sound like overkill, but thought given to such matters with effective implementation can cut down the time spent on basic tasks. In this context it is worth thinking about the Premack principle, mentioned earlier, and working out your hierarchy of preferred tasks. You can then make sure you reward yourself with a preferred task after a less preferred one. As an interesting aside on this, a recent survey showed that the first thing many people do on reaching the office in the morning is have a cup of coffee. This is presumably a preferred activity. Evidence suggests that spending time on a less preferred task before having coffee would get that task done more quickly than if it were left until later in the day. As this anecdote implies, all these techniques can be used equally well at work as at home.

The Behavioural Approach Applied to Meetings

Some managers in one of the organizations in which we have worked decided to apply the approach to improving performance in their meetings. They did this with some success. It is necessary to compile an agreed checklist of appropriate behaviours. This could include:

- Everyone present within three minutes of specified start time.
- Everyone has all the necessary papers.
- Everyone has read all the necessary papers.
- Only one person speaking at any time.
- No interruptions from outside sources.
- No deviation from relevant business.
- Action points clear, agreed and recorded.

Just as with the safety programme, it is first necessary to establish the baseline for appropriate behaviours and then to set a goal for improvement. Observations are then continued for a predetermined period, during which, it is hoped, the percentage of appropriate behaviours will approach or even exceed the goal.

It should by now be clear that the behaviour-based approach can be applied to almost any situation, and as a final example we include in Figure 7.2 a checklist we once produced for changing a wheel on a car, following a puncture.

A good quality job is one that is done safely, gives the end result of a correctly inflated tyre and is done within ten minutes.

	Conform	Not Conform	Not Relevant
Warning triangle erected 15 metres from car	____	____	____
Warning flashers operating	____	____	____
Tool kit complete (including pressure gauge)	____	____	____
Wheels chocked before jacking starts	____	____	____
Wheel nuts loosened before jacking starts	____	____	____
Wheel nuts stored in suitable container	____	____	____
Jack extended sufficiently to fit inflated tyre	____	____	____
Spare has legal tread	____	____	____
Spare inflated to correct pressure	____	____	____
Replacement tyre pressure matched to its partner	____	____	____
Punctured tyre securely stored	____	____	____
All tools securely stored	____	____	____
No tools remaining on road	____	____	____
Remove wheel chocks	____	____	____
Switch flashers off	____	____	____
Drive away	____	____	____

Figure 7.2 *The behavioural approach in an everyday situation: checklist for changing a car wheel following a puncture*

Conclusion

It is our firm contention that an organization can always take and use a good idea so long as it is based on:

1 Sound theory and an understanding of the principles underlying that theory.
2 The understanding that the approach must be adapted to fit the needs of the particular organization.

Our selling point, as far as we have one, is that it is usually not effective to buy an off-the-shelf package and expect it to work in the same way as a carefully tailored process. Any innovation must be tailored to fit with current work practices and systems so as not to be too disruptive, if it is to gain acceptance by the workforce. Furthermore, commitment to the new process, and its enduring continuation over a long period of time, will only occur if both management and workforce feel that they have ownership of that process. We feel that the behavioural approach to safety improvement fulfils these conditions and facilitates a move away from the traditional 'boom and bust' strategies. The approach we are recommending will keep the process renewed and up-to-date. It does, however, require effort, in the same way that effort is required to keep plant and equipment functioning effectively. The key to success is in ensuring employee involvement in the decisions that affect their safety and performance in the workplace.

In our experience the time and effort required to successfully set up and initiate a behaviour-based approach to continuous safety improvement brings benefits that go far beyond simply creating a safer working environment. As we have shown, in this chapter, the potential exists to use the model to increase participation and involvement for *all* employees in all aspects of their work, bringing benefits in which everyone will share. We have introduced this approach into many different work environments and, each time, we learn something new from the people we train and help to set up the system. We continue to be impressed by the ideas put forward to improve the basic model. It is clear that a wealth of untapped knowledge and experience exists out there in the workplace. This is a valuable resource, if only the management and organizational structures will allow it to be released. We believe that the behaviour-based approach provides a vehicle for this to happen.

Appendix A

OUTLINE PROGRAMME FOR THE OBSERVER TRAINING COURSE

DAY ONE

0900   Course Introduction

    Personal Introductions
    Course Objectives
    Hopes and Fears

0930   The Theory of the Behaviour-based Approach to Safety

Presentation of basic theory underlying the behavioural approach sufficient to enable participants to understand the reason for the procedures.

1200   Developing a Checklist

A practical exercise to develop understanding and skill in designing checklists.

1230   Lunch

1300   Technical Aspects of Being an Observer

Instruction in how to complete a checklist and discussion of potential problems of being an observer.

1330   Practice in Completing Checklists

Participants make a trial run at completing the checklist in their own work context.

1400   Scoring a Checklist

Instruction on how to score the checklist and establish the baseline measure.

1500   Identification and Modification of Checklist Items

Participants review and update the checklists they are going to use.

1630   Finish

DAY TWO

0900   Practical Aspects of Being an Observer

Presentation and exercises on the interpersonal aspects of being an observer. Includes:

> Effects of cooperation and competition
> Staying OK
> Dependence theory

1100   Feedback

Presentation on goal-setting, feedback charts and interpersonal aspects of giving feedback.

1200   Summary, Questions, Hopes and Fears Revisited

Overview and review of hopes and fears expressed at the beginning of the programme.

1230   Close

Appendix B

A SEMI-STRUCTURED INTERVIEW SCHEDULE WITH SOME QUESTION EXAMPLES

- Introductions/setting the scene/icebreaker session.
- Guarantee of confidentiality.
- Ask for permission to take notes; offer to show the notes to the interviewee at the end of the session.
- 'Do you want to ask me anything before we begin?'
- Prompt: 'Did you receive the letter from XXXX? What are your first impressions about the approach?'

Section 1 Understanding of 'quality' as a concept

A1 What do you understand by (the word) 'quality'?
A2 Is quality important? If so, why?
A3 Can you give an example of what would be good quality (in your work)? Why and how is it good?
A4 Can you give an example of what would be bad quality (in your work)? Why and how is it bad?
A5 Do you know the costs/consequences of poor quality for the plant/ department or in your job?

Section 2 The management of quality and potential outcomes

B1 Does the company have a policy on quality?
B2 What has been done to manage/improve quality performance? In what ways has this changed what happens?
B3 Have attitudes to quality changed in the past 2 years? If so, how?
B4 Do attitudes to quality differ within the company – how/why?

Section 3 The customers

C1 Who are your customers?
C2 How do you assess what the customers think about quality of product?
C3 What do your customers think of the company/department/section in terms of quality? Has this changed? If so, how and why?

C5 Have customers' views changed as a result of the previous quality improvement programmes?

Section 4   Responsibility for quality

D1 When quality goes wrong for your customers, whose fault is this, and why?

D2 What aspects of quality are in your hands? To what extent can you influence the quality of the product/service?

D3 What do you do to achieve quality? What extra could you do? What prevents you from doing this?

D4 Does it matter if you do a 'quality' job? In what ways?

D5 Is quality included in the way you are appraised?

D6 What happens when quality goes wrong; who gets blamed/punished? Are you praised or given feedback when quality is good?

Section 5   Others

E1 What do you consider to be the major barrier to improving quality in the company/this department or shift etc.?

E2 What do you think should be done to improve quality?

E3 If you were the boss, what would you do to improve quality?

E4 We are helping to develop a quality programme in this plant – is there anything you think we should know about or think about when we do this?

Section 6   Additional questions for management

F1 What is the role of quality in the company's overall strategy?

F2 Who is driving quality?

F3 What is driving quality?

F4 Is the management of quality consistent with the achievement of short-term goals?

F5 Does BS 5750/ISO 9000 form part of the company's quality policy? If so, why? What effect will this have?

F6 To what extent is quality about changing organizational culture? If so, how are you going about this?

F7 If the quality programme works 'perfectly', how would the plant look in 5 years time?

- 'Do you have any further questions for me or do you wish to add any other comments?'
- Explanation of what will happen to the information obtained and how it will be used to design the questionnaire.

Thank the interviewee and close the interview.

REFERENCES

Adam, E.E., Jr (1975) Behaviour modification in quality control. *Academy of Management Journal*, 18: 662–678.

Adam, E.E., Jr and Scott, W.E., Jr (1971) The application of behavioural conditioning procedures to the problems of quality control. *Academy of Management Journal*, 14 (June): 175–193.

Ballard, R. (1993) Keeping a TABS on safety. *Journal of Occupational Safety and Health*, 23 (12): 41–43.

Ballard, R. (1995) Changing behaviour, improving safety. *Health, Safety and Environment Bulletin*, IRS No. 229.

Caplan Inquiry (1997) 'Piper Alpha – two men blamed', cited in *Safety and Health Practitioner*, October, p. 2.

Chin, R. and Benne, K.D. (1976) General strategies for effecting changes in human systems. In W.G. Bennis, K.D. Benne, R. Chin and K.E. Carey (eds), *The Planning of Change*, 3rd edn. New York: Holt, Rinehart and Winston.

Cooper, M.D., Phillips, R.A., Sutherland, V.J. and Makin, P.J. (1994) Reducing accidents using goal setting and feedback: a field study. *Journal of Occupational and Organizational Psychology*, 67: 219–240.

Cox, C.J. and Makin, P. (1994) Overcoming dependence with contingency contracting. *Leadership and Organization Development Journal*, 15(5): 21–26.

Cullen, The Hon. Lord (1990) *Report of the Public Inquiry into the Piper Alpha Disaster*. London: HMSO.

Dale, B. and Cooper, C.L. (1992) *Total Quality and Human Resources: an Executive Guide*. Oxford: Blackwell.

Donald, I. (1994) Measuring psychological factors in safety. *The Health and Safety Practitioner*, March, pp. 26–29.

Duff, A.R., Robertson, I.T., Phillips, R.A. and Cooper, M.D. (1994) Improving safety by the modification of behaviour. *Construction Management and Economics*, 12: 67–78.

Eagley, A.H. and Chaiken, S. (1993) *The Psychology of Attitudes*. Fort Worth, TX: Harcourt Brace.

Ernst, F. (1971) The OK Corral: the grid for 'get on with'. *Transactional Analysis Journal*, 1: 4.

Feigenbaum, A.V. (1986) Quality improvement and learning in productive systems. *Management Science*, October, pp. 1305–1315.

Fennell, D. (1998) *King's Cross Underground Station Fire, 18 November 1987*. London: Department of Transport Investigations Report.

Festinger, L. (1962) Cognitive dissonance. *Scientific American*, October.

Fowler, A. (1998) How to cut absenteeism. *People Management*, 4 (1): 44–45.

French, J.R.P. and Raven, B.H. (1959) The bases of social power. In D. Cartwright (ed.), *Studies in Social Power*. Ann Arbor: University of Michigan Press.

Hackman, J.R. and Oldham, G.R. (1975) *The Job Diagnostic Survey*. New Haven, CT: Yale University, Department of Administrative Sciences Technical Report No. 4.

Hall, B. (1983) *OBM in Multiple Business Environments*. New York: Haworth Press.

Harrington, H.J. (1987) *The Improvement Process. How America's Leading Companies Improve Quality*. New York: McGraw Hill.

Harrison, R. (1987) *Organization Culture and Quality of Service*. London: Association for Management Education and Development.

Hayes, R.H. and Clark, K.B. (1985) Explaining observed productivity differences between plants: implications for operations research. *Interfaces*, November–December, pp. 3–14.

Health and Safety Commission (1997) *HSC Plan of Work 1997/98: Summary*. October, C65, misc. 085. London: HSE Books.

Health and Safety Commission (1997) *HSC Annual Report 1996/97: Summary*. November, C80. Misc. 094. London: HSE Books.

Heinrich, H.W. (1950) *Industrial Accident Prevention: A Scientific Approach*. New York: McGraw Hill.

HSE (Health and Safety Executive) (1992) *Successful Health and Safety Management*. HS(G)65. London: HMSO.

Kaeter, M. (1992) Personalising performance. *Quality*, April, pp. 36–41.

Kahneman, D. and Tversky, A. (1984) Choices, values and frames. *American Psychologist*, 39: 341–350.

Kim, J.S. and Hamner, W.C. (1976) Effect of performance feedback and goal setting on productivity and satisfaction in an organizational setting. *Journal of Applied Psychology*, 61(1): 48–57.

Komaki, J. (1998) *Leadership from an Operant Perspective*. London: Routledge.

Komaki, J., Barwick, K.D. and Scott, L.R. (1978) A behavioural approach to occupational safety: pinpointing and reinforcing safe performance in a food manufacturing plant. *Journal of Applied Psychology*, 63 (4): 434–445.

Komaki, J., Waddell, W.M. and Pearce, M.G. (1977) The applied behavior analysis approach and individual employees: improving performance in two small businesses. *Organizational Behavior and Human Performance*, 19: 337–352.

Krause, T.R. and Finley, R.M. (1993) Safety and continuous improvement. Two sides of the same coin. *The Safety & Health Practitioner*, September, pp. 19–22.

Landau, J.C. (1993) The impact of a change in an attendance control system on absenteeism and tardiness. *Journal of Organizational Behavior Management*, 13(2): 51–70.

LaPiere, R.T. (1934) Attitudes vs actions. *Social Forces*, 13: 203–237. Quoted in Sabini, J. (1995) *Social Psychology*. New York: W.W. Norton.

Lewin, K. (1947) Group decisions and social change. In T. Newcomb and E. Hartley (eds), *Readings in Social Psychology*. New York: Holt, Rinehart and Winston.

Lievegoed, B.C.J. (1973) *The Developing Organization*. London: Methuen.

Likert, R. (1961) *New Patterns of Management*. New York: McGraw Hill, p. 113.

Locke, E.A. and Latham, G.P. (1991) *A Theory of Task Setting and Goal Performance*. New York: Prentice–Hall.

Luthans, F. and Kreitner, R. (1975) *Organizational Behaviour Modification*. Glenview, IL: Scott, Foresman & Co.

Luthans, F. and Kreitner, R. (1985) *Organizational Behaviour Modification and Beyond. An Operant and Social Learning Approach*. Glenview, IL: Scott, Foresman & Co.

McAfee, R.B. and Winn, A.R. (1989) The use of creative feedback to enhance workplace safety: a critique of the literature. *Journal of Safety Research*, 20: 7–19.

McCarthy, M. (1978) Decreasing the incidence of 'high bobbins' in a textile spinning department through a group feedback procedure. *Journal of Organizational Behaviour Management*, 1: 150–154.

Makin, P.J. and Hoyle, D.J. (1993) The Premack Principle: Professional Engineers. *Leadership and Organizational Development Journal*, 14(1): 16–21.

Makin, P.J. and Sutherland, V.J. (1991) A fatal inversion. *Journal of Occupational Safety and Health*, 21(11): 40–42.

Peters, T.J. and Waterman, R.H. (1982) *In Search of Excellence*. New York: Harper and Row.

Porter, L.W., Steers, R.M., Mowday, R.T. and Boulian, P.V. (1974) Organizational commitment, job satisfaction and turnover among psychiatric technicians. *Journal of Applied Psychology*, 59 (5): 603–609.

Reporting of Injuries, Diseases and Dangerous Occurrences Regulations (RIDDOR) (1995) London: HMSO.

Royal Society for the Prevention of Accidents (RoSPA) (1997) *Occupational Safety and Health Bulletin*. November, p. 16 ISP/P396-1.

Rousseau, D.M. (1988) The construction of climate in organizational research. In C.L. Cooper and I.T. Robertson (eds), *International Review of Industrial and Organizational Psychology*. Chichester: Wiley.

Sabini, J. (1995) *Social Psychology*. New York: W.W. Norton.

Safety Statistics Bulletin 1996/97. C40, misc. 083. London: HSE Books.

Schonberger, R.J. (1982) *Japanese Manufacturing Techniques: Nine Hidden Lessons in Simplicity*. New York: Free Press.

Sell, R.G. (1977) What does safety propaganda do for safety? A review. *Applied Ergonomics*, 8(4): 203–214.

Simon, H.A. (1990) Invariants of human behavior. *Annual Review of Psychology*, 41: 1–19.

Squires, F.H. (1980) *Successful Quality Management*. Columbus, OH: Hitchcock Publications.

Sutherland, V.J. and Makin, P. (1993) An attitude problem? *Journal of Occupational Safety and Health*, July, pp. 44–47.

Sutherland, V.J., Makin, P.J., Bright, K. and Cox, C.J. (1995) Quality behaviour for quality organisations. *Leadership and Organization Development Journal*, 16 (6): 16–26.

Sutherland, V.J., Makin, P.J., Phillips, R. and Cooper, M.D. (1993) An attitude problem. *Journal of Occupational Safety and Health*, 23(7): 44–47.

Sykes, P. (1996) Empowerment. Nobody said it would be easy. *Works Management*, March, pp. 37–41.

Tarrants, W.E. (1980) *The Measurement of Safety Performance*. London: Garland.

Warr, P., Cook, J. and Wall, T. (1979) Scales for the measurement of some work attitudes and aspects of psychological well-being. *Journal of Occupational Psychology*, 52: 129–148.

Wheelwright, S. (1981) 'Japan – where operations really are strategic', *Harvard Business Review*, July–August, pp. 67–76.

Zohar, D. and Fussfeld, N. (1981) A systems approach to organizational behavior modification: theoretical considerations and empirical evidence. *International Review of Applied Psychology*, 30: 491–505.

INDEX